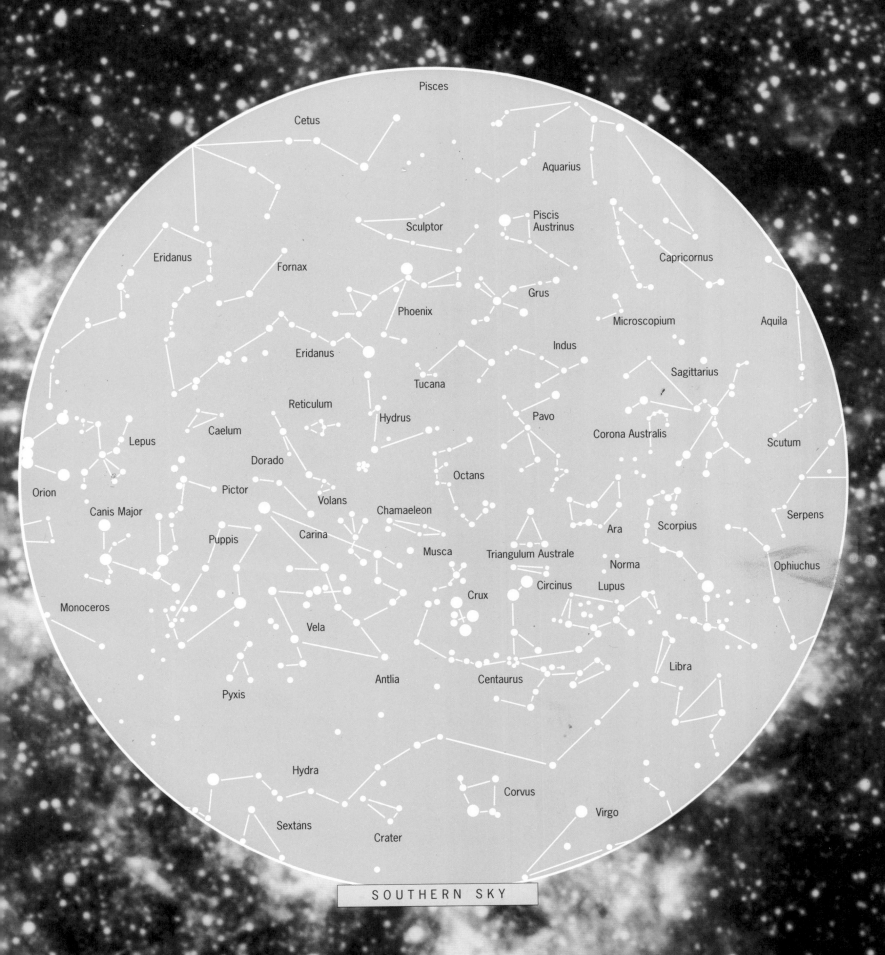

SOUTHERN SKY

The ATLAS of Space EXPLORATION

The ATLAS of Space EXPLORATION

Tim Furniss

THE ATLAS OF SPACE EXPLORATION

Z-Publishing Ltd.
7-11 St. Johns Hill
London SW11 1TN
United Kingdom

This edition published in North America
2002 by Michael Friedman Publishing Group, Inc.
Originally published by Z-Publishing Ltd. © 2001

ISBN 1-58663-346-5
10 9 8 7 6 5 4 3 2 1

Created and produced by
Fire Publishing Limited
1 Torriano Mews, London NW5 2RZ
United Kingdom

Project Director
Nicholas Bevan

Editor
Quentin Daniel

Art Editor
Bob Burroughs

Editorial Services
Sam Merrell, Susan Wheatley

Picture Researchers
Tim Furniss, Vanessa Fletcher

Illustrations by
Julian Baker, Colin Woodman

Color origination by
Vistascan, Singapore

Printed in Spain

Contents

Voyages into the unknown

To our remote ancestors, the night sky was the domain of the mysterious powers that shaped the destinies of humankind. In their eyes, only gods would have had the ability to travel above the Earth's surface. Yet today, as space stations and satellites orbit our planet and as the space shuttle undertakes regular missions, the idea of space travel no longer seems extraordinary. In fact, many people even believe that one day it may be as common as air travel is now. Such achievements were the culmination of centuries of curiosity about the universe and our place in it, and were made possible by a century of scientific and technological progress. But above all, it was the courage of the early space travelers, willing to risk their lives like the mariners of old on the great oceans, that enabled human beings to take his first steps beyond the "final frontier."

Above:
The space shuttle in orbit deploys cargo from its payload bay.

Above:
Astronaut works on crane outside the International Space Station.

The stargazers

See also:
- **Space observation** *p. 10*
- **The Hubble Space Telescope** *p. 76*

STONEHENGE

Stonehenge in the west of England (below) is one of the world's oldest observatories. Constructed between 1800 and 1400 B.C., the astronomical observations made in this massive circle of stones were probably used to calculate calendar dates for religious rites. The outer ring of 56 stones is the oldest part. Together with a "heel stone" some distance away, these were aligned to receive the rays of the rising Sun at the summer solstice. A second, inner ring of stones was raised about 100 years afterward in 1700 B.C. The stones lying on top of pairs of uprights were still later additions.

FROM THE EARLIEST TIMES, humans have looked to the night sky to explain our presence on Earth. Not surprisingly, many ancient observers studied the Moon and stars for religious reasons. The first truly scientific observations were made by the Babylonians in about 2000 B.C., and their successors included a number of Greek astronomers whose views were well ahead of their time. But it was not until the 15th and 16th centuries, when Copernicus and Galileo published their ideas, that there began to be broad acceptance of the fact that the Earth is round and orbits the Sun.

Below:
The cosmic plan of the Greek astronomer Ptolemy, with Earth at the center of the universe.

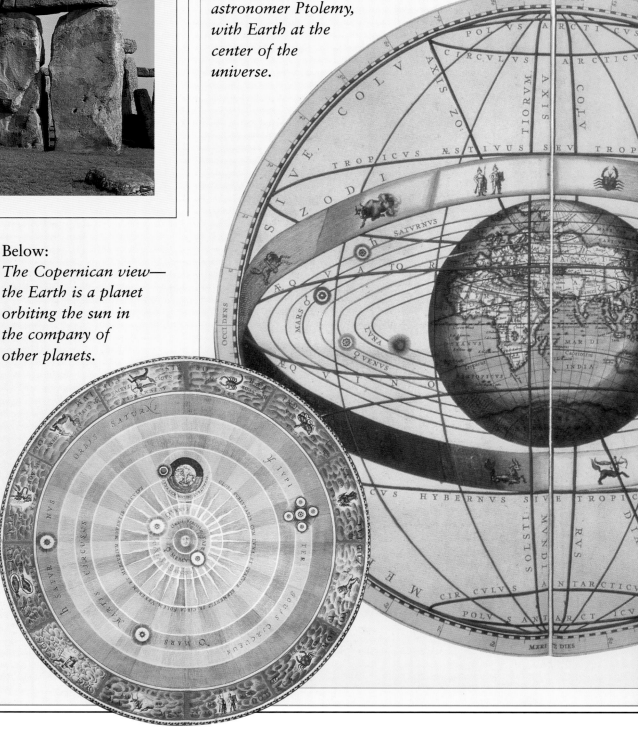

Below:
The Copernican view—the Earth is a planet orbiting the sun in the company of other planets.

Above:
Polish astronomer Nicolaus Copernicus, whose "On the Revolution of the Celestial Spheres" was published in 1543.

EARLY TELESCOPES

The Italian scientist Galileo Galilei (left) is credited with being the first person to study space through a telescope. In 1609 he was able to examine the surface of the Moon, the rings of Saturn, and Jupiter and its moons. A much more powerful "reflecting" telescope (below), which used mirrors to collect light, was developed by the English scientist Sir Isaac Newton about 60 years later. Later on, astronomers such as William Herschel built large reflectors through which the very structure of our galaxy could be studied. These instruments were the forerunners of today's giant telescopes.

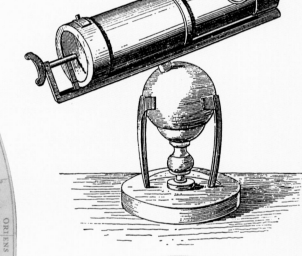

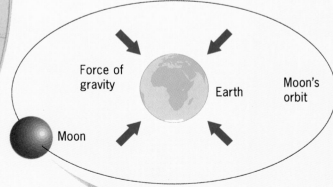

Above:
The Earth and Moon in Newton's theory of gravity. Newton explained that a large body in space will exert gravitational "pull" on a smaller one.

Force of gravity

Earth

Moon's orbit

Moon

Moon hurtles into space without gravity

THE PATHFINDERS

Though they had only crude wooden measuring instruments, the ancient Egyptians were extremely skillful at pinpointing the apparent positions of the stars and lined up the pyramids in accordance with them. The Great Pyramid, built in about 2700 B.C., was aligned to what was then the pole star, Thuban in the constellation of the Dragon.

THE EGYPTIANS

The Danish astronomer Tycho Brahe (1546–1601) is regarded as the greatest observer of the pretelescopic era. He made detailed records of the positions of the planets and saw two supernova explosions. The latter were proof to him that the universe is not a fixed system but is constantly changing.

TYCHO BRAHE

Johannes Kepler (1571–1630) was Brahe's assistant. His own major contribution to astronomy was to set out the laws of planetary motion, one of which was that the planets did not travel around the Sun in a circle but in an ellipse. Kepler also built his own telescope and wrote a book about comets.

JOHANNES KEPLER

The British astronomer Sir William Herschel (1738–1822) discovered the planet Uranus using a self-built reflecting telescope in 1781. In 1785, on the command of King George III, he made the world's largest telescope. It had a reflecting mirror 4 feet (1.2 m) in diameter—a size not exceeded until 1845.

WILLIAM HERSCHEL

Space observation

See also:
- **The stargazers** p. 8
- **The Hubble Space Telescope** p. 76

ANTIPODEAN VIEWS

Australia is an ideal place for observatories because of its generally clear atmosphere. The Siding Spring Observatory (below), sited on a 3,821-foot (1,165 m) mountain at the edge of Warrumbungle National Park, New South Wales, houses the 12.8-foot (3.9 m) diameter *Anglo-Australian Telescope* (AAT). The observatory's dome is 164 feet (50 m) high, opens to expose the telescope when it is operational, and can be rotated to allow the massive instrument to point at selected targets. The observatory also contains a wide variety of other, smaller telescopes.

Right:
An astronomer rides in the "prime-focus cage" of the Anglo-Australian Telescope (above) as it takes photographs with exposure times of 60 to 90 minutes.

THE DEVELOPMENT of large telescopes in the early 20th century revolutionized human knowledge. Such instruments brought home to observers the vastness of the universe—and our own tiny place in it. By the 1930s, radio telescopes were being used to detect "radio noise" from the depths of space, while the 1970s saw the launch of the first space observatories to study distant planets and stars in different wavelengths such as ultraviolet, infrared, gamma ray, and X ray. But Earth-based observatories continue to provide us with vital information.

Above:
The Pleiades star cluster in a photograph taken by the Royal Observatory, Edinburgh, Scotland.

Right:
A cross-section of a typical computer-controlled observatory telescope.

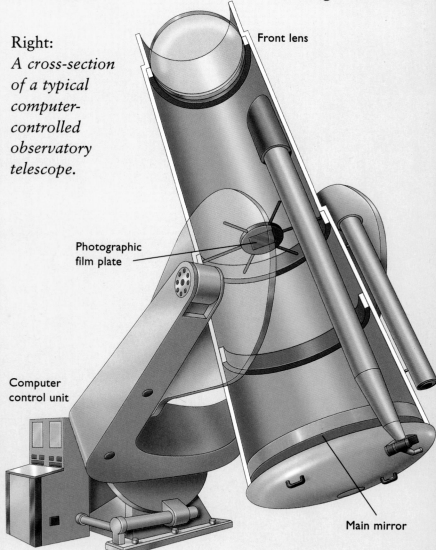

Front lens

Photographic film plate

Computer control unit

Main mirror

THE GREAT LISTENER

Slung like a hammock in a natural hollow between hills, the Arecibo Observatory in Puerto Rico (right) is the world's largest radio telescope. The 1,000-foot (305 m) diameter dish is very sensitive and can pick up radio waves emitted from very remote objects in space. Though the telescope itself cannot be steered, the sky from 43° north to 6° south can be observed by moving "feed aerials" supported on three towers on its rim.

Above:
An Anglo-Australian Telescope image showing the globular star cluster 47 Tucanae, which contains several million stars. In the Cone Nebula (left), new stars are born.

Left:
An engineer at the Anglo-Australian Telescope's controls monitors the instrument's second-by-second observations, as "slow exposure" photography reveals stars (right) moving around the celestial south pole of the night sky above.

Rocket science

See also:
- **The rocket age** *p. 14*
- **The space race** *p. 16*

BALLOON PRINCIPLE

Rocket propulsion or "thrust" is created by the force of the exhaust of burnt fuel passing through a nozzle. It is just as if you inflated a balloon and let go of the mouthpiece (left). The air passing under pressure through the balloon's nozzle creates thrust, which in turn sends the balloon flying uncontrollably through the air. However, rocket thrust must be carefully controlled and the rocket kept on course by a guidance system. By adjustments to the nozzle's angle, the rocket can also be steered.

Left:
The "father" of rocket science—the Russian Konstantin Tsiolkovsky, pictured in 1930. Tsiolkovsky laid down many of the theories exploited by later space programs.

A ROCKET IS a vehicle propelled by the pressure of exhaust created by an engine burning fuel. It is as yet the only means of space travel. Basic rocket design has not changed much since the early days of development in World War II, when the German *V2* was used as a missile to attack Allied targets. A basic rocket engine needs a solid or liquid fuel and an oxidizer, which produces the combustion. Most rockets are in the form of "boosters" that carry spacecraft or satellites into Earth's orbit. Once their mission is completed and their fuel runs out, they fall back to Earth.

Above:
An Atlas Centaur rocket, burning liquid oxygen and kerosene, lifts off from Cape Canaveral.

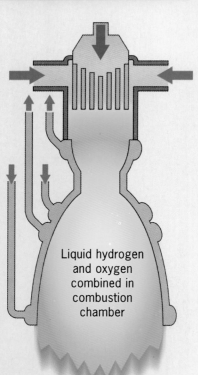

Liquid hydrogen and oxygen combined in combustion chamber

Hot exhaust gases provide thrust

Above:
The space shuttle's main engines and rocket boosters seen close-up during launch, at a distance (right), and in section (left).

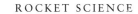

V2 INVENTOR

Wernher von Braun (right) was the German rocket engineer who helped develop the *V2* missile during World War II. Packed with high explosives, the missile was launched with often devastating effect at over 3,000 Allied targets. After the war von Braun went to the U.S. The *V2* versions he developed there became the basis of the U.S. space program.

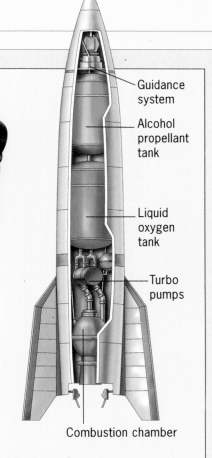

Guidance system

Alcohol propellant tank

Liquid oxygen tank

Turbo pumps

Combustion chamber

Left:
Cape Canaveral, 1950—the first launch of the 46-foot (14 m) tall modified V2 rocket, seen also in section (right). It was fueled by liquid oxygen-alcohol.

Below:
The U.S. Athena solid rocket booster carrying the Lunar Prospector into space. A section of the shuttle's solid rocket booster (right).

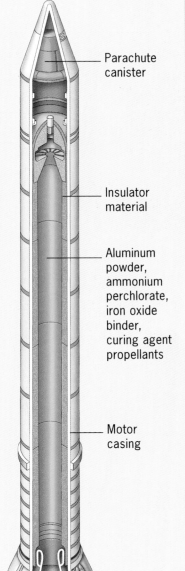

Parachute canister

Insulator material

Aluminum powder, ammonium perchlorate, iron oxide binder, curing agent propellants

Motor casing

Nozzle

The rocket age

See also:
- **Rocket science** p. 12
- **The space race** p. 16

SPACE COMMERCE

Rocket-launching has become a major business concern, with commercial launcher companies charging between $50 and $100 million to carry a customer's satellite into geostationary transfer orbit. The U.S., Russia, China, and Europe all promote various commercial rockets to this end (other rockets are operated for government launches of military satellites). The leader in the market is the European company Arianespace, which operates a fleet of *Ariane 4* and *5* rockets.

Left:

A Chinese Long March 3B rocket carries a Hong Kong-operated communications satellite into space.

TODAY, NUMEROUS COUNTRIES have space programs and employ rockets for a variety of purposes. In fact there are about 50 launches each year, most being for communications satellites (though there are also some for Earth observation, scientific, or military satellites). A rocket needs a number of stages to reach space, each stage falling away as its propellants run out. Two stages thrust the satellite into the upper atmosphere, while a third stage puts it into orbit. Final Earth orbit is usually attained about 10 minutes after the rocket lifts off.

Left:

Europe's powerful Ariane 5 lifts off from Kourou, South America. A typical three-stage rocket (right) has:
(1) boosters
(2) first stage
(3) payload shrouds
(4) second stage
(5) payload
(6) instrument section
(7) third stage.

ROCKET DATA

Rocket	Operator	Weight of cargo
Ariane 5	Arianespace (Europe)	14,994 lb (6,800 kg)
Delta IV Medium	Boeing (U.S.)	14,553 lb (6,680 kg)
Sea Launch	Sea Launch (U.S./International)	11,023 lb (5,000 kg)
Long March 3B	CGWIC (China)	11,023 lb (5,000 kg)
Ariane 44L	Arianespace (Europe)	10,805 lb (4,900 kg)
Proton K	ILS (U.S./Russia)	10,143 lb (4,600 kg)
Atlas IIIB	ILS (U.S./Russia)	9,923 lb (4,500 kg)
H2 Alpha Plus	Rocket Systems (Japan)	8,818 lb (4,000 kg)

LAUNCH SHOTS

Photographs of rocket launches are taken by automatic cameras (above), as in addition to the heat generated during a launch, the sound waves would kill a photographer standing nearby. Positioned at different locations on the rocket gantries, these cameras are heavily armored to counter the effects of the blast. Close-up film cameras are also used by engineers to monitor the performance of the rocket. These can help scientists to trace the cause of any accident.

Above:
The insulation panels surrounding the second stage of a European Ariane 44L rocket fall away during launch, while the first stage of a U.S. Atlas II (right) sheds ice formed by its liquid oxygen.

Above:
Engineers are dwarfed by a Titan 4 at Cape Canaveral, as a small navigation satellite, attached to the third stage of Delta 2 (right), is prepared in its payload shroud.

The Space race

See also:
- **Space pioneers** *p. 22*
- **Man on the Moon** *p. 48*

SPUTNIK 1

Sputnik 1 (below) was a 176-pound (80 kg) aluminum sphere with four aerials. During its 92-day orbit of the Earth in 1957, the radio transmitter inside sent out "bleep, bleep" signals that could be picked up all over the world. Its mission completed, *Sputnik* reentered the Earth's atmosphere and burnt up. Few people outside the Soviet Union believed the country had the technology to put the first satellite in space. In reply, the U.S. government ordered top rocket scientist Wernher von Braun to push ahead with an American satellite. *Explorer 1* was launched the following year.

T HE SPACE AGE began during the 1950s, a time of worldwide tension when nuclear war between the Soviet Union and the U.S. seemed a real possibility. When the Soviet Union launched the first satellite, *Sputnik 1*, on October 4, 1957, it caused the Americans great concern. *Sputnik 1*'s rocket was the first intercontinental missile, capable of carrying a nuclear warhead to the U.S. within minutes. So the ensuing race to dominate space, when the two countries competed in a series of satellite launches, started as much for military as for purely scientific reasons.

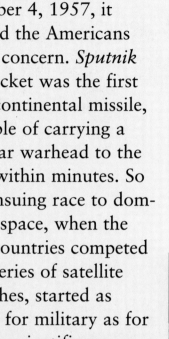

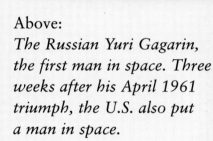

Above:
The Russian Yuri Gagarin, the first man in space. Three weeks after his April 1961 triumph, the U.S. also put a man in space.

Left:
An American artist's impression of the lunar surface, made in the early days of the space race when the U.S. answer to Sputnik, Explorer 1 (below), was launched on a Jupiter C *rocket.*

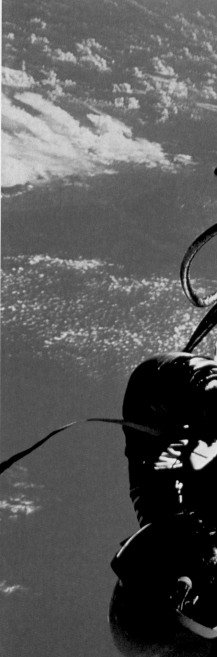

KENNEDY'S COMMITMENT

In 1961, U.S. President John F. Kennedy (below) made a commitment to put an American on the moon by the end of the decade. This sent a message to the Soviet Union that the resources of the world's richest nation would be poured into the space program and stressed that a landing on the Moon would be the ultimate test of a country's technological progress in space exploration. This was achieved by the *Apollo 11* team in 1969, but sadly, Kennedy himself was not to see the goal achieved: He was assassinated in Texas on November 22, 1963.

Above:
Saturn 5, *the massive rocket booster needed to blast the U.S. Apollo modules into Earth's orbit.*

Above:
The race is won—U.S. astronaut Buzz Aldrin steps onto the lunar surface on July 21, 1969.

Left:
A U.S. Gemini 4 astronaut spacewalks— part of the intensifying competition between the U.S. and the Soviet Union.

Satellites in orbit

See also:
- **Types of satellite** p. 20
- **Space stations** p. 26

SOLAR POWER

The life expectancy of many of the early satellites was quite limited, because they depended on internal batteries. Today, by contrast, most satellites are powered by solar cells mounted on deployable winglike arrays (below). Each mirrorlike, silicon solar cell retains the Sun's energy, which is then converted into electrical power. When the satellite is in the Earth's shadow, it reverts to battery power.

POWERED BY solar arrays, bristling with antennae, and packed with hundreds of sensitive components such as thermostats and signal processors, some 2,500 satellites orbit the Earth. The orbit a satellite adopts depends on the particular job it has to do. Being the easiest to reach, low Earth orbits are used by space stations and the *Hubble Space Telescope*. Mapping satellites are often in highly inclined polar orbits, and communications satellites in geostationary orbit, traveling at the same speed as the Earth rotates. Lastly, astronomy satellites in elliptical orbit travel low over the Earth at first, then go far out to space, hanging for a long time at the farthest point of their "loops."

Below:
The satellite has small internal fuel tanks for the rocket thrusters and engines needed for repositioning in space.

Below:
Satellites are launched into orbits with differing inclinations to the equator. A 90° inclination orbit would take the satellite over the poles.

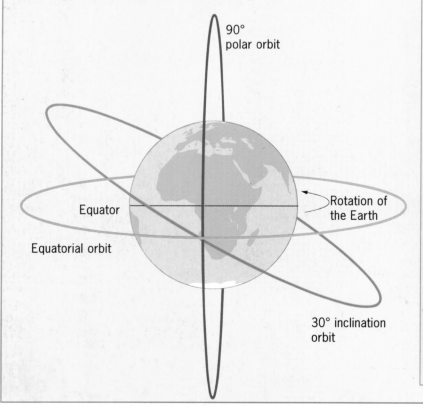

90°
polar orbit

Equator

Rotation of the Earth

Equatorial orbit

30° inclination orbit

SPACE DEBRIS

Debris often ends up in orbit along with the satellites. This often includes the spent upper stage of the satellite's rocket (left) along with insulation material and even a few nuts and bolts. If it still has propellant in its tanks, the rocket stage presents a real danger. The propellant will overpressurize and an explosion may occur. This may cause highly expensive damage to other satellites and lead to yet more space debris. At the moment there are at least 6,000 pieces of "space junk" orbiting the Earth.

Above and below:
Communications satellites, which contain a maze of electrical wires and are fitted with huge, sensitive antennae.

Right:
A satellite's rocket motor, used to fire the craft into the correct orbit.

Below:
Satellites enter geostationary orbit in three stages: low Earth orbit, geostationary transfer orbit, and, following a rocket engine firing, the correct orbit itself.

Circular
low Earth orbit

Launch

Earth

Geostationary
transfer orbit

Geostationary
orbit

Types of satellites

See also:
- **The space race** *p. 16*
- **Satellites in orbit** *p. 18*

TELSTAR

The 170-pound (77 kg) satellite *Telstar 1* (below) was launched in July 1962 and transmitted the first live TV pictures from the U.S. to Europe. A 31-inch (0.8 m) sphere developed by the American Telephone and Telegraph Company, *Telstar* transmitted signals to both U.S. and European ground stations when its orbit took it over the Atlantic.

SATELLITES ARE an essential part of our technological age. Whenever we turn on the television, make a mobile telephone call, listen to the weather forecast, or fly in an airliner, a satellite is likely to be involved in some part of the process. Satellites for monitoring the weather and environment, for communications, espionage, and navigation— all transmit information daily to receivers across the world. Despite the huge cost of building satellites and putting them into space, there are large long-term profits to be gained from them, and over 50 are launched every year. Earth's orbit is becoming an increasingly cluttered place.

Above:
A European Remote Sensing satellite, which monitors Earth's resources from a low polar orbit using a radar instrument.

Left:
The highly sensitive infrared telescope on this military satellite can detect a missile launch anywhere in the world.

SATELLITE FACTS

October 1957–June 2001
- Launched: 5,394
- Russian satellites: 3,150
- U.S. satellites: 1,543
- Japanese satellites: 82
- Countries that have launched satellites with national rockets: 16. They include China, the United Kingdom, Brazil, France, and various consortiums.

Right:
Infra-Red Space Observatory, *a European astronomical observatory in Earth's orbit.*

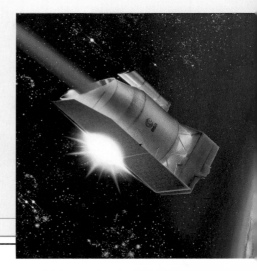

RETRIEVABLE SATELLITES

Once the majority of satellites are in orbit, they remain there until they are replaced by more modern spacecraft. Others in low Earth orbit, on the other hand, eventually reenter the atmosphere and burn up. But some scientific or astronomy satellites are deployed on a strictly short-term basis. These satellites are carried by the space shuttle into orbit, given a brief independent flight, and then recaptured by the shuttle's robot arm.

Above:
A Spartan free-flying satellite, held at the end of the shuttle's remote manipulator robot arm before being deployed for its independent flight in space.

Right:
The Odyssey Launch Platform with the Sea Launch Vehicle erect and ready for blastoff.

Left:
Communications satellites are equipped with dish antennae which receive and redirect signals to specific areas of the Earth.

OCEAN-BASED LAUNCHES

The Thuraya-1 mobile communications satellite, bearing the heaviest commercial payload to date, was launched into orbit from a sea platform in October 2000. Weighing 11,260 pounds (5,108 kg), it was the work of a multinational team from the U.S., Russia, and the Ukraine. It has a 12-year lifespan and orbits the Earth at a height of 22,212 miles (35,768 km).

Space pioneers

See also:
- **The space race** p. 16
- **Man on the Moon** p. 48

ANIMALS IN SPACE

Research animals paved the way for manned spaceflight. At first small animals, such as mice, were launched in rockets flying short, up-and-down suborbital flights. Then the dog, Laika, (left), strapped inside the Russian *Sputnik 2*, was launched into Earth's orbit on November 3, 1957. Unfortunately, the satellite was not designed to be recovered and Laika died a week later. Two other Russian dogs were more fortunate when their satellite was recovered in 1960. In the U.S., chimpanzees were sometimes used in experimental flights. Both Ham (right), who made the first suborbital Mercury flight, and Enos, who made the first fully orbital one, were recovered after

Right:
The U.S. Mercury Redstone 3 mission blasts off on May 5, 1961, taking astronaut Alan Shepard on his suborbital flight.

TODAY, MANNED SPACEFLIGHT seems almost commonplace. But between 1961 and 1966, millions of people around the world followed the stories of the first 24 manned missions with a mixture of wonder and awe. Spaceflight then was a comparatively crude science—astronauts were blasted into the atmosphere on the top of ballistic missiles —and full of potential hazards. The bravery of the first 30 space travelers cannot be underestimated. It is also quite remarkable that, though some missions experienced problems, no astronaut was killed until 1967.

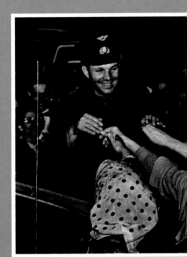

Above:
Lieutenant Yuri Gagarin of the Soviet Air Force is greeted by crowds following his Earth orbit on April 12, 1961.

SPACE FIRSTS

- April 12, 1961: Yuri Gagarin is first man in space
- May 5, 1961: Alan Shepard is first American on suborbital flight
- August 6, 1961: Soviet cosmonaut flies in orbit for a day
- June 16, 1963: First woman in space launched by Soviet Union
- October 12, 1964: Soviet Union launches three men into orbit
- March 18, 1965: First spacewalk made by a Soviet cosmonaut
- December 15, 1965: U.S. astronauts complete first space rendezvous
- March 16, 1966: First space docking achieved by U.S. astronauts
- April 16, 2001: First "tourist" on *International Space Station*

John Glenn's space patch
for his Mercury flight

THE MERCURY SEVEN

In 1959, the U.S. space agency NASA selected seven air force pilots to head up their manned Mercury space program. These men had to go through rigorous training, including endurance and weightlessness tests, in order to become America's first astronauts. Such was the intense interest in space travel in the West, the "Mercury Seven" were famous even before they set foot in their space capsules. By contrast, the names of the cosmonauts in the Soviet Union's space program were kept secret until after their launch.

Alan Shepard

America's first man in space was launched aboard *Freedom 7* on May 5, 1961. His suborbital flight lasted just 15 minutes and 28 seconds. In 1971, Alan Shepard became the fifth man on the Moon. He died at age 74 on July 21, 1998.

Gus Grissom

Grissom was launched aboard *Liberty Bell 7* on July 21, 1961. His suborbital flight lasted 15 minutes and 37 seconds. He also led the first manned Gemini mission in 1965. He was killed in a fire during a prelaunch test for *Apollo 1* in 1967.

John Glenn

America's first man in orbit was launched aboard *Friendship 7* on February 20, 1962. His three orbits of the Earth lasted four hours and 55 minutes. He later became a U.S. senator. In October 1998, at the age of 77, he went back into space aboard the space shuttle.

Scott Carpenter

Carpenter was launched aboard *Aurora 7* on May 24, 1962. His three orbits of the Earth lasted four hours and 56 minutes. Unfortunately, he missed his splashdown target by 250 miles (402 km)—and never made another space flight.

Wally Schirra

Schirra was launched aboard *Sigma 7* on October 3, 1962. His six orbits of the Earth lasted nine hours and 13 minutes. In December 1965 he commanded *Gemini 6*, and in October 1968 he flew the Apollo project test flight, *Apollo 7*.

Gordon Cooper

Cooper was launched aboard *Faith 7* on May 15, 1963. His mission lasted just under a day. Later, in August 1965, he commanded *Gemini 5* on an eight-day mission. He also trained to make an Apollo Moon landing but was not selected.

Deke Slayton

Slayton was barred from taking part in Project Mercury in 1962 by a slight heart defect. But in July 1975, at the age of 51, he eventually flew as docking module pilot of the combined Apollo-Soyuz mission. He died on June 13, 1993.

The space shuttle

See also:
- **Space stations** *p. 26*
- **Living and working in space** *p. 28*

THE SHUTTLE IN SPACE

The space shuttle *Discovery* in Earth's orbit (below), photographed in 1993 by the unmanned satellite *Orpheus*. Carrying an ultraviolet astronomy telescope, the satellite had just been deployed from the spacecraft's payload bay and was designed to operate on its own in orbit. It was later retrieved by the *Discovery* crew using the robot arm or remote manipulator system (RMS), seen here cocked at an angle to the payload bay. The RMS is used on most shuttle missions to deploy and retrieve satellites and move spacewalking astronauts, who stand in special footholds at the end of the arm, from one part of the payload bay to another.

Above:
The Shuttle Endeavor *blasts skyward on an 11-day mission in October 2000.*

LAUNCHED IN 1981, the space shuttle was the world's first reusable spaceship. Since then it has flown almost 130 missions in Earth's orbit, its crew conducting valuable scientific research—such as mapping the Earth's surface in minute detail—and repairing or deploying satellites. The orbiter itself has three main liquid propellant engines. At blastoff these are fueled by a massive external tank that falls away and disintegrates in the atmosphere once the shuttle has reached orbital altitude. The thrust of the main engines during the first two minutes is increased by two solid rocket boosters.

THE BIGGEST CREW

The eight members of the largest crew to fly on a shuttle mission (above), pose for a camera during their mission in October 1985. The first four shuttle test flights carried a mere two test pilots. By the fifth mission, however, two specialist astronaut crew members had been added. Shuttle missions today normally carry between five and seven crew members and may also include nonastronauts such as scientists and technicians. The shuttle now ferries crews and supplies to and from the *International Space Station (ISS)*.

Above:
A fish-eye lens on a 35-mm camera was used to record this panorama of the Earth's horizon, with the cargo bay of the shuttle Atlantis in the foreground.

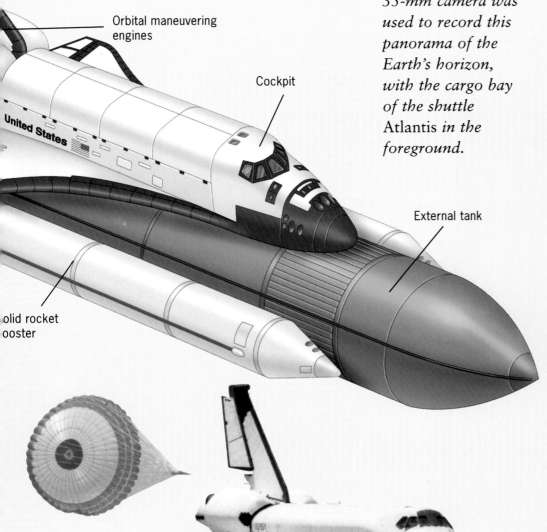

Orbital maneuvering engines

Cockpit

United States

External tank

olid rocket
ooster

Above:
The shuttle on the launch pad at Kennedy Space Center, Florida, and touching down (below) after its mission, and an illustration showing it mounted on its fuel tank (left).

Space stations

See also:
- **The space shuttle** p. 24
- **Living and working in space** p. 28

FUTURE VISIONS

The first designs of space stations (see below) were highly imaginative, even fantastic visions. Drawn up well before the space age itself got under way, they were based on the assumption that getting the many necessary parts into space for assembly would be as easy as transporting cargo from one country to another by air. But to launch space station materials into orbit is a complicated and risky business, as both the Soviet and U.S. programs demonstrated. It is also hugely expensive. By the time the new *International Space Station* is completed in 2004, for example, it is reckoned that it will have cost at least $90 billion.

THE FIRST space station was *Salyut 1*, a single pressurized module launched by the Soviet Union in 1971. Over the next 15 years, several Salyut stations were manned by cosmonauts brought up by Soyuz craft. The U.S. program, on the other hand, was slower to get off the ground. The first *Skylab* space station was launched in 1973 and was manned by three astronaut crews in succession. But it was the Soviet Union's launch of *Mir* in 1986 that initiated the real advance in space station technology. However, the launch of the *International Space Station* (*ISS*) in 1997 marked the most successful breakthrough in establishing a permanent base in space. International crews have manned *ISS* continually, and new missions bring ever-more advanced technology to the base to monitor the universe and improve communications.

Above:
The Soviet Salyut *space statio in orbit, with its winglike sola arrays deployed.*

Below:
The ISS contrasted against the cloud-covered Earth and the inky darkness of space.

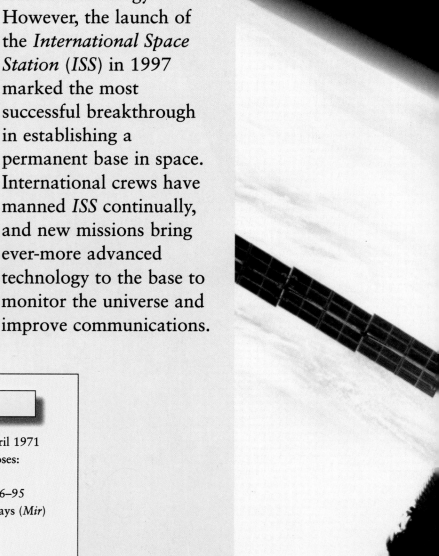

Above:
The U.S. Skylab *space station. Its huge laboratory was made from a spare Apollo Saturn rocket stage.*

STATION FACTS

- First space station: *Salyut 1*, launched April 1971
- First space station used for military purposes: *Salyut 3*, launched 1974
- Construction in space of *Mir* station: 1986–95
- Record time spent in space station: 436 days (*Mir*)
- Launch of first *ISS* component: 1998
- First crew on *ISS*: year 2000

MIR IN SPACE

Like any extremely complex piece of machinery, *Mir* (left) had its problems and had to undergo numerous repairs. While these are a routine part of space operations, the impression was sometimes given of unreliability. Still, *Mir* was a remarkable space station, housing over 30 crews. Toward the end of its 15-year career, the NASA space shuttle docked with it several times, providing stepping stones toward the establishment of the *ISS*.

Below:
An American astronaut handles the boom of a Russian crane during a six-hour EVA (extravehicular activity) in May 2000.

Right:
A head-on view of the ISS taken by a crew member on space shuttle Discovery prior to docking. It shows the supply ship Progress attached to the service module Zvezda.

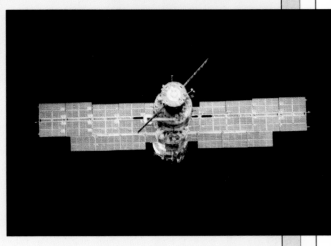

Living and working in space

See also:
- **The space shuttle** *p. 24*
- **Space stations** *p. 26*

THE SMALLEST ROOM

Visits to the bathroom in the early days of space exploration were a fairly primitive business: The astronauts had to defecate into plastic bags that they took back to Earth for scientific analysis. By contrast, the space shuttle's bathroom (left) has a seat like that on an airliner. But there the resemblance ends. The astronaut must urinate into a special receptacle attached to a hose, while a vacuum pump ensures that solid waste does not float away in the shuttle's near-zero gravity!

COMPARED WITH the early days of manned spaceflight, conditions in space today are positively luxurious. The space shuttle and space stations have food galleys, sleeping bunks, bathrooms, washing facilities, and even in-flight entertainment like films and music. Nonetheless, every spaceflight involves an enormous amount of hard work for the astronauts or cosmonauts involved. The shorter the mission, the busier, and space travelers must be prepared to deal with the unexpected, such as making an emergency space walk to repair something on the craft's exterior.

Below:
Clad in his bulky pressure suit, a space shuttle astronaut works in the payload bay of the orbiter.

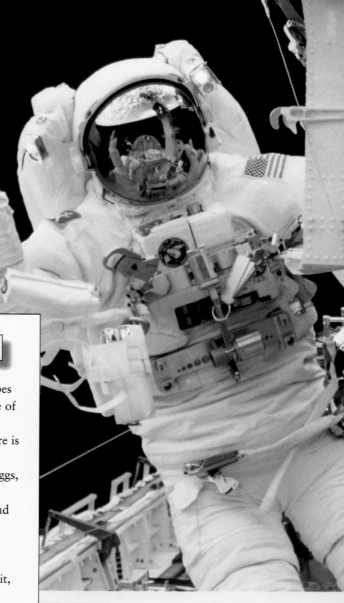

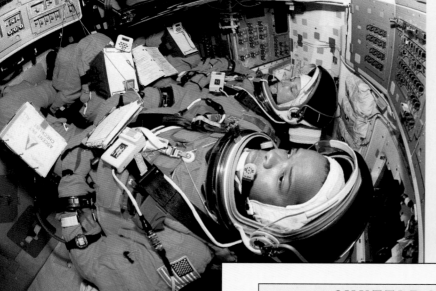

SHUTTLE MENU

Above:
Countdown to blastoff—two mission specialists strapped in for launch on the flight deck of the space shuttle. The commander and pilot are seated above them.

Early space travelers ate liquidized food squeezed from tubes or dried bite-sized pieces. Today the astronaut has a choice of freeze-dried, irradiated, or thermostabilized food. Space Shuttle crews select their daily menus before the flight. Here is a typical example:
- Breakfast—Bran flakes, coffee, orange juice, scrambled eggs, and breakfast roll.
- Lunch—Grape juice, cauliflower cheese, strawberries, and pecan cookies.
- Dinner—Shrimp cocktail, beef stroganoff with noodles and green beans, and pears.
- Available all day—Coffee, tea, juices, water, biscuits, fruit, and snack bars.

CRAMPED CONDITIONS

One of the biggest problems encountered by *Mir* cosmonauts (left) was simply keeping the station tidy. Regular ferry trips by spacecraft to and from future stations should enable crew members to get rid of garbage as and when it is created. But overall, such stations should be far more spacious than the close confines of either the space shuttle or *Mir*. Private quarters for each crew member will be just one of the major improvements in living conditions.

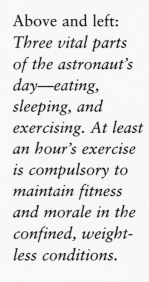

Above and left: *Three vital parts of the astronaut's day—eating, sleeping, and exercising. At least an hour's exercise is compulsory to maintain fitness and morale in the confined, weightless conditions.*

Discovering new worlds

T HE EARTH IS ONE PLANET in a solar system of planets and moons orbiting the Sun, our local star. The other planets have long intrigued astronomers. Well into this century there was speculation that intelligent life might exist on Mars or Venus, and it took the expeditions of unmanned space probes to discredit the idea. But what was discovered on these and other planets proved almost as fascinating. For whether flying past, orbiting, or actually landing on the surfaces of these strange worlds, spacecraft have added enormously to scientific knowledge of our neighbors in the solar system. The information they have sent back not only confirms the extraordinary variety of even this small corner of the universe, it emphasizes the special status of Earth as a haven of life in an otherwise desolate void.

Above:
The spacecraft Galileo *is deployed in orbit by the space shuttle.*

Above:
U.S. Mariner 2 *was launched in May 1961, reaching Venus in December 1962.*

The solar system

See also:
- **The Milky Way** p. 80
- **The universe** p. 84

THE PLANETS

Planet	Average distance from the Sun	Diameter
Mercury	36 mil. mi (57.9 mil. km)	3,030 mi./4,880 km
Venus	67 mil. mi (108.2 mil. km)	7,514 mi./12,100 km
Earth	93 mil. mi (149.6 mil. km)	7,921 mi./12,756 km
Mars	142 mil. mi (227.9 mil. km)	4,215 mi./6,787 km
Jupiter	483 mil. mi (778.3 mil. km)	88,679 mi./142,800 km
Saturn	886 mil. mi (1,427 mil. km)	74,893 mi./120,600 km
Uranus	1,782 mil. mi (2,870 mil. km)	32,168 mi./51,800 km
Neptune	2,793 mil. mi (4,497 mil. km)	30,740 mi./49,500 km
Pluto	3,664 mil. mi (5,900 mil. km)	1,428 mi./2,300 km

Below:
The inner ring of planets, which could hardly be more varied. While Mercury has no atmosphere, the atmosphere on Venus is 90 times more dense than on Earth. The atmosphere on Mars is thin and consists mainly of carbon dioxide.

Mercury

Venus

NINE PLANETS and their moons, along with asteroids, comets, and meteors are under the sway of the Sun's gravity. Together they form just one system of matter surrounding one average-sized star in the Milky Way. Yet the sheer size of our solar system is enough to overwhelm the imagination. In terms of relative scale, if the Sun were a big watermelon, the Earth would be a pea 98 feet (30 m) away; Pluto would be a grain of sugar 0.9 miles (1.5 km) distant. Moreover, over 99 percent of the solar system's mass is concentrated in the Sun itself.

Above:
The sole haven of life in the solar system—the Earth, seen here with the Moon. Like all the planets apart from Pluto, the Earth orbits the Sun on the so-called ecliptic plane (below). Pluto's orbit is more diagonal to the Sun.

Below:
The near and the far—Mercury orbits as close as 28.5 million miles (45.9 million km) to the Sun, while Mars can orbit 155 million miles (249.1 million km) away from it.

Earth

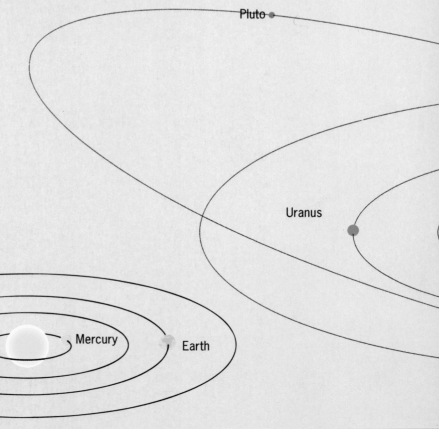

Pluto

Uranus

Mars

Mercury

Earth

Venus

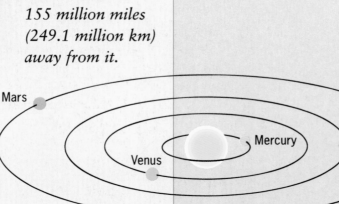

Mars

Jupiter

Saturn

Uranus

Neptune

Neptune

Pluto

Above:
The outer ring of planets. Jupiter, Saturn, Neptune, and Uranus have a large number of moons, while the farthest planet, Pluto, is scarcely bigger than a moon itself. Pluto takes over 1,000 times longer to orbit the Sun than Mercury.

Saturn

...iter

The Sun

See also:
- **The solar system** *p. 32*
- **Stars and constellations** *p. 78*

SOHO VISION

By normal, "visible" light, the Sun looks tranquil enough. Space-borne instruments of different wavelengths, however, have revealed the truth of our star's turbulence. These images taken by the *European Solar and Heliospheric Observatory* (SOHO) not only show giant bursts of flame erupting from the Sun's circumference but have also registered the hottest regions on the Sun's disc as specks of brightness.

THE SUN is just one, relatively small star among millions in our galaxy, the Milky Way. Small though it is for a star, the Sun is still 109 times bigger than the Earth, and is essentially a great ball of light- and heat-generating gases. These gases are produced by a continuous controlled nuclear reaction or fusion. Every second 700 million tons of hydrogen fuse to form helium, which in turn releases enormous energy in the form of heat and light. Without the Sun, the Earth would be like the Moon—cold, dark, and utterly lifeless.

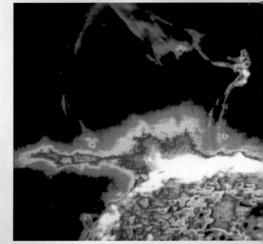

Above:
Hot gas erupts from the Sun in this spectro-heliograph image from Skylab.

Left:
The space observatory SOHO, *launched in 1995 to conduct the most intensive observations of the Sun and the solar wind ever made.*

SUN DATA

- Sunlight takes 8 minutes 17 seconds to reach the Earth
- The Earth orbits the Sun at a distance varying between 93 and 94 million miles (149.6–152 million km)
- The Sun has a diameter of 865,000 miles (1,392,000 km)
- Photosphere temperature ranges from 7,800–16,200 °F (4,300–9,000 °C)
- The chromosphere contains gases as hot as 1.8 million °F (1 million °C)
- The corona can be as hot as 7.2 million °F (4 million °C)

SUN FACTS

- Storms in the photosphere form sunspots
- Sunspots send out solar flares
- The upper level of the photosphere is called the stormy chromosphere
- Above the chromosphere is the corona, a halo of still hotter gases
- The outer layer of the corona, that streams away from the Sun, is called the solar wind

THE AURORA PHENOMENA

Auroras are extraordinary cosmic light displays—pulsing bands of color high in the Earth's atmosphere—that can be seen only in the far north and south of the globe. They are created when charged particles from the solar wind become trapped in the Earth's magnetic field. The particles spiral in lines of the magnetic field and interact with gases in the upper atmosphere. The aurora borealis occurs in northern latitudes, the aurora australis in southern ones.

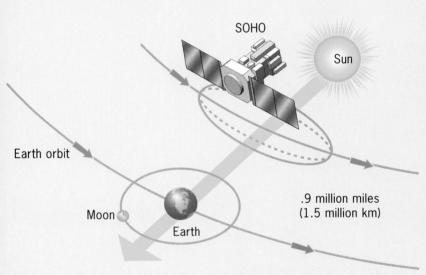

SOHO

Sun

Earth orbit

Moon

Earth

.9 million miles
(1.5 million km)

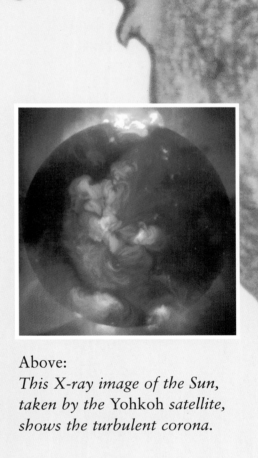

Above:
This X-ray image of the Sun, taken by the Yohkoh *satellite, shows the turbulent corona.*

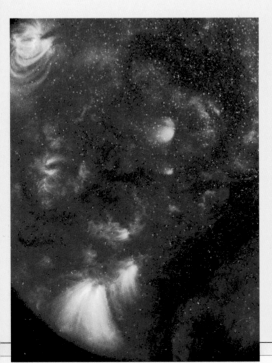

Above:
SOHO's *orbit lets it observe the Sun continuously.*

Left:
A solar flare, seen in ultraviolet by the Trace *satellite.*

Right:
The Sun is formed of the core, the photosphere, chromosphere, and corona.

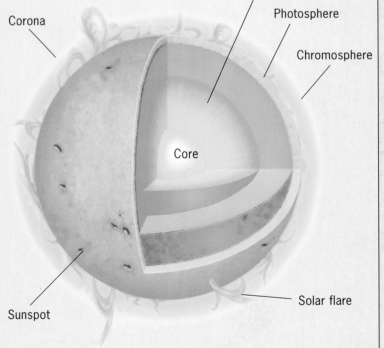

Corona

Radiative zone

Photosphere

Chromosphere

Core

Sunspot

Solar flare

Mercury

See also:
- **The solar system** *p. 32*
- **The Moon** *p. 44*

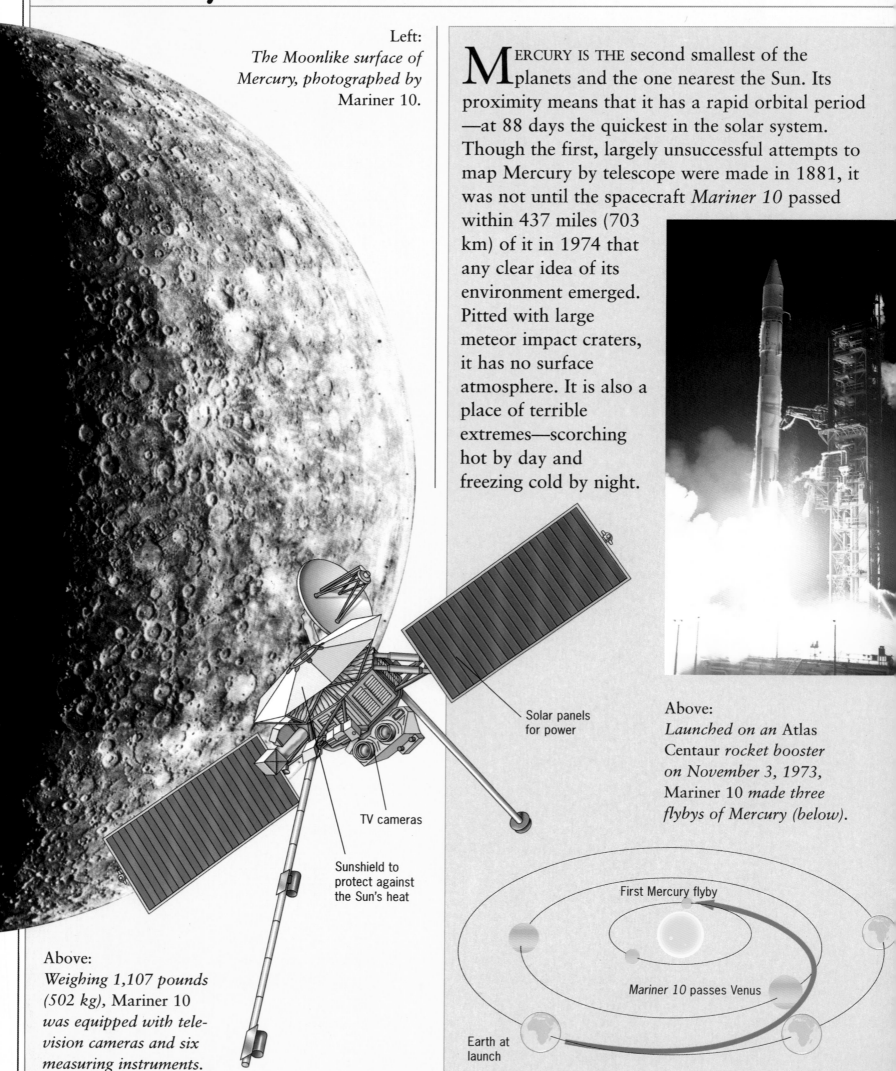

Left:
The Moonlike surface of Mercury, photographed by Mariner 10.

Mercury is the second smallest of the planets and the one nearest the Sun. Its proximity means that it has a rapid orbital period —at 88 days the quickest in the solar system. Though the first, largely unsuccessful attempts to map Mercury by telescope were made in 1881, it was not until the spacecraft *Mariner 10* passed within 437 miles (703 km) of it in 1974 that any clear idea of its environment emerged. Pitted with large meteor impact craters, it has no surface atmosphere. It is also a place of terrible extremes—scorching hot by day and freezing cold by night.

Solar panels for power

TV cameras

Sunshield to protect against the Sun's heat

Above:
Launched on an Atlas Centaur *rocket booster on November 3, 1973,* Mariner 10 *made three flybys of Mercury (below).*

First Mercury flyby

Mariner 10 passes Venus

Earth at launch

Above:
Weighing 1,107 pounds (502 kg), Mariner 10 was equipped with television cameras and six measuring instruments.

Earth and Mercury compared

Left:
The largest meteor impact crater on Mercury has a diameter of 124 miles (200 km). The planet also has a huge 807-mile (1,300 km) impact basin named Caloris.

GOING TO MERCURY

Mariner 10's first stop in the solar system was Venus, the gravity of which was used to "slingshot" the spacecraft on its path to Mercury. *Mariner 10*'s first encounter with the planet took place on March 29, 1974, when it came to within 437 miles (703 km) of the surface. Two further flybys on September 21, 1974 and March 16, 1975 brought the craft ever closer to the planet. No other spacecraft have been sent to Mercury since, but *BepiColombo* is scheduled for launch in late 2002 on a 2.5-year journey to Mercury. It will spend a year in orbit studying the planet.

Below:
Mercury's slow rotation in relation to its rapid orbit of the Sun means that its daytime and nighttime are extraordinarily drawn out: each lasts 176 Earth days.

Mercury

Sun

MERCURY DATA

● Max. distance from Sun: 43.3 mil. mi (69.7 mil. km)
● Min. distance from Sun: 26 mil. mi (45.9 mil. km)
● Diameter: 3,030 miles (4,880 km)
● Orbits the Sun: once every 87.96 days
● Rotation period: 58.64 days
● Speed through space: 29.7 miles per sec (47.87 kps)

Thin rocky crust

Mantle

Iron core containing 80 percent of planet's mass

Above:
With its large iron core, Mercury is one of the densest planets in the solar system. Its surface (below) is a desert whose temperature veers from 788 °F to -292 °F (420 °C to -180 °C).

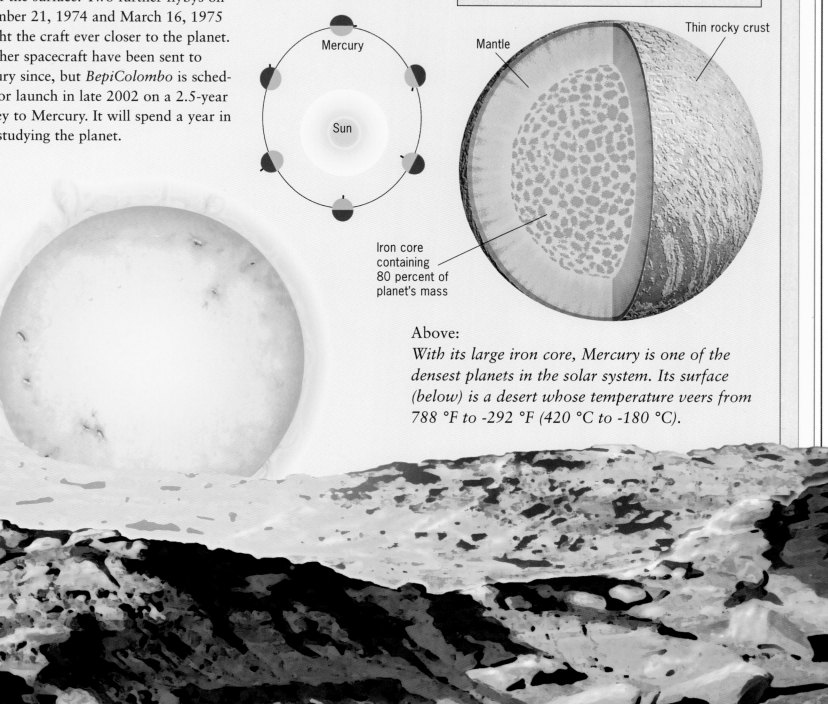

Venus

See also:
- **The solar system** *p. 32*
- **Mapping Venus** *p. 40*

PIONEER PROGRAM

At Cape Canaveral in 1978 the first of two *Pioneer-Venus* spacecraft (below) was launched aboard an *Atlas Centaur* booster on a mission to orbit Venus and deposit four probes on its surface. The orbiter arrived on December 7 and conducted an intensive scientific survey of the planet. It was joined two days later by the *Multiprobe* spacecraft. The latter deployed one large and three small probes, each of which made parachute landings onto the surface. The probes encountered 224 miles per hour (360 kph) winds in the upper atmosphere but little surface wind.

VENUS IS the brightest object in the night sky after the Moon. This is because the thick cloud covering the planet makes it highly reflective to sunlight. Before the days of space exploration, scientists speculated that its relative proximity to the Sun would make Venus a humid world of tropical jungle. This view was soon overturned by spacecraft, which proved instead that the planet's environment is one of the most extreme in the solar system. Not only is the surface temperature 887 °F (475 °C), but atmospheric pressure is 90 times that of Earth and it rains sulfuric acid.

Above:
An image of the Venus surface sent back in 1982 by a Russian Venera lander (left). The Venera program's first success came in 1967 (below).

Above:
US Mariner 2, the first spacecraft to explore the planet Venus (December 1962).

VENUS DATA

- Max. distance from Sun: 68 mil. mi (109 mil. km)
- Min. distance from Sun: 67 mil. mi (107.4 mil. km)
- Diameter: 7,514 miles (12,100 km)
- Orbits the Sun: once every 244 days
- Rotation period: 243 days
- Speed through space: 6.4 miles per sec (10.36 kps)
- Surface temperature: 887 °F (475 °C)
- Atmosphere: 98 percent carbon dioxide

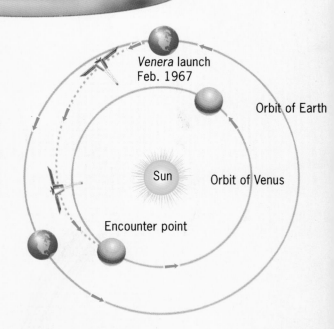

Venera launch
Feb. 1967

Orbit of Earth

Sun

Orbit of Venus

Encounter point

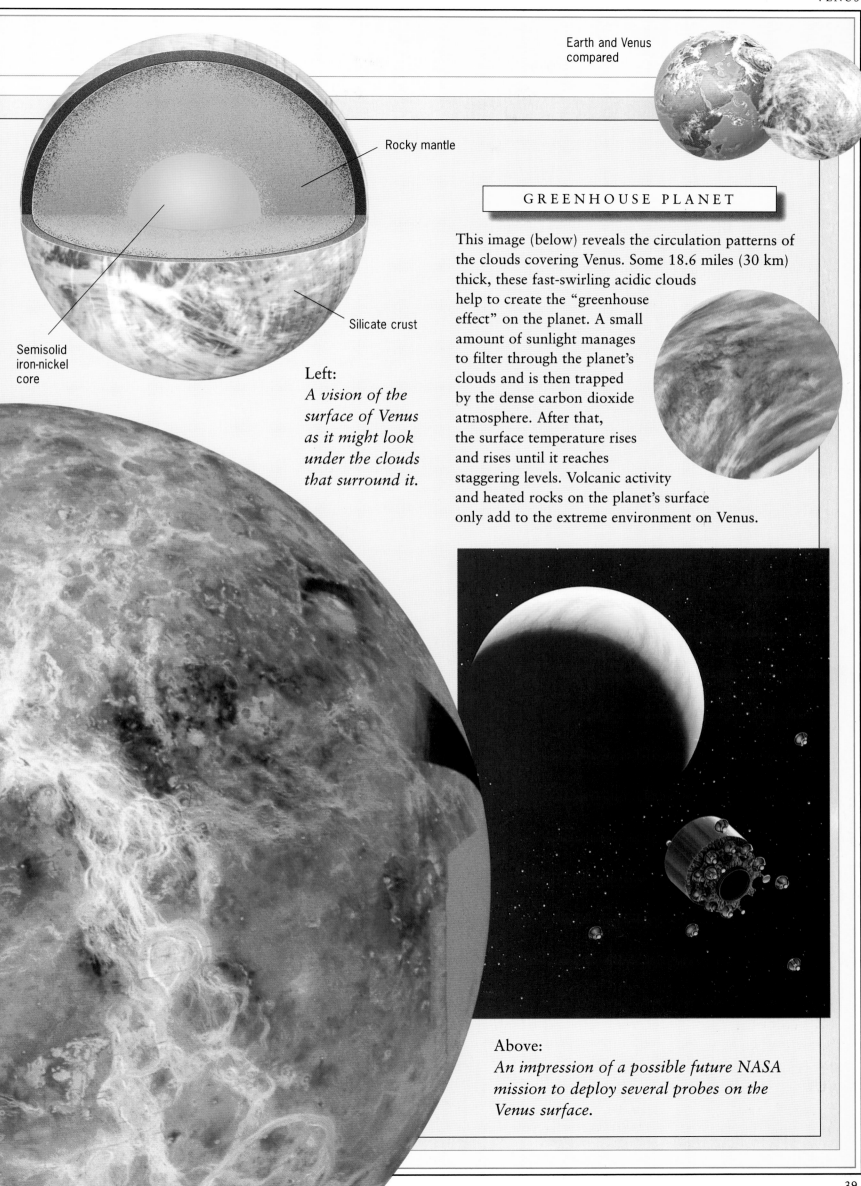

Earth and Venus
compared

Rocky mantle

Silicate crust

Semisolid
iron-nickel
core

Left:
*A vision of the
surface of Venus
as it might look
under the clouds
that surround it.*

GREENHOUSE PLANET

This image (below) reveals the circulation patterns of
the clouds covering Venus. Some 18.6 miles (30 km)
thick, these fast-swirling acidic clouds
help to create the "greenhouse
effect" on the planet. A small
amount of sunlight manages
to filter through the planet's
clouds and is then trapped
by the dense carbon dioxide
atmosphere. After that,
the surface temperature rises
and rises until it reaches
staggering levels. Volcanic activity
and heated rocks on the planet's surface
only add to the extreme environment on Venus.

Above:
*An impression of a possible future NASA
mission to deploy several probes on the
Venus surface.*

Mapping Venus

See also:
- **The solar system** *p. 32*
- **Venus** *p. 38*

UNVEILING VENUS

The spacecraft *Magellan* headed for its rendezvous with Venus in August 1990. Near the planet, a retro rocket was fired and *Magellan* entered its orbit over the poles. This heralded three years of operations (effectively four 243-day cycles), during which the spacecraft used its radar (see below) to map almost the whole planet. After this, *Magellan* performed a series of "aerobraking" maneuvers to demonstrate how the atmosphere could be used to slow down a spacecraft. Eventually, in October 1994, *Magellan* plunged toward Venus and burned up.

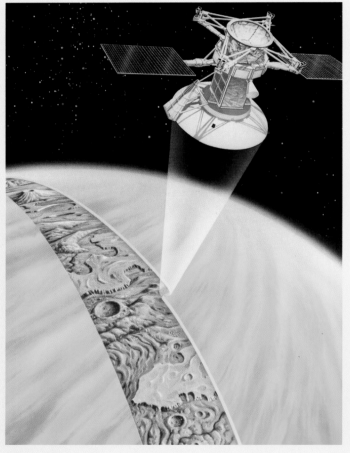

EARLY EXPLORATION of Venus was made by spacecraft using cameras that were unable to "see" through the planet's thick cloud cover. Since cameras could not pierce the clouds over the planet, new technology was required. The answer lay in radar imaging. Radar signals are sent through the atmosphere to the surface, and the signals "bounced" off objects and back up to the space-craft. By measuring the differences in these signals, a terrain profile is built up. This idea worked. The first radar pictures were returned by the U.S. *Pioneer-Venus* orbiter in 1979. Two Russian orbiters, *Venera 15* and *16*, followed in 1983. Finally, in 1989, the radar-mapper *Magellan* was sent to the planet.

Above:
Magellan being deployed by the space shuttle Atlantis *in Earth's orbit.*

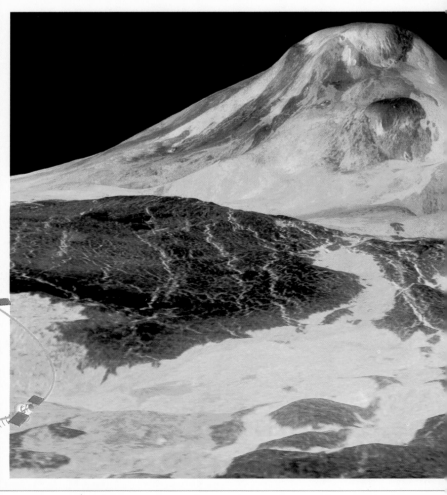

Above:
Magellan in the course of its radar surveys of 90 percent of Venus.

Orbital path

Magellan

Earth

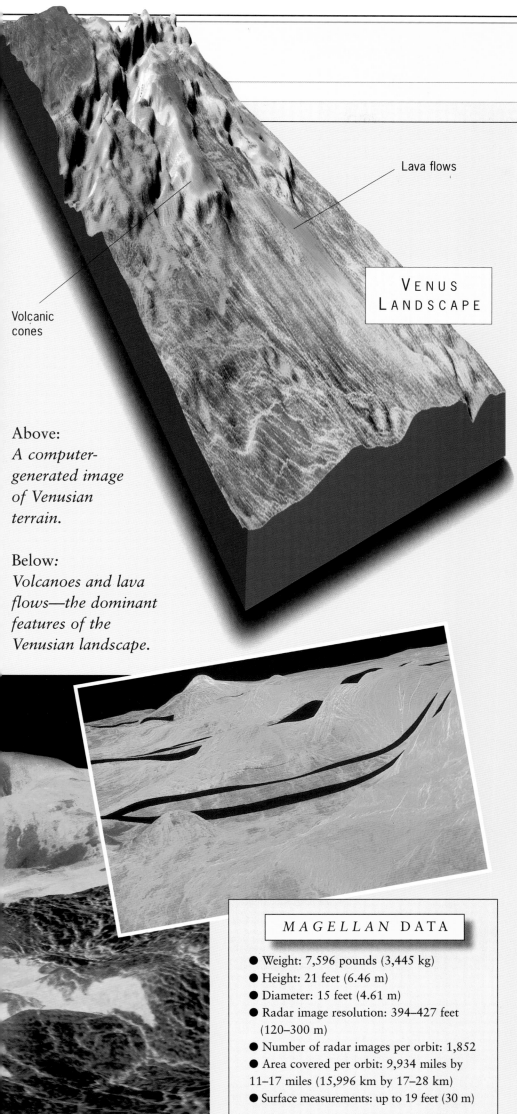

Lava flows

Volcanic
cones

VENUS
LANDSCAPE

Above:
*A computer-
generated image
of Venusian
terrain.*

Below:
*Volcanoes and lava
flows—the dominant
features of the
Venusian landscape.*

SURFACE FEATURES

Much of the chaotic terrain of Venus appears to have been caused by recent volcanic activity. The low number of impact craters indicates that volcanoes have replaced the original features of the planet. *Magellan* images also reveal a range of other features, including bright ejecta (thrown-up material) near the impact craters, fractured plains, and the U.S.-sized Ishtar Terra highlands.

The various linear patterns, or ridge belts, in the relatively recent lava flows of Venus are signs of complex volcanic activity on the planet's surface. The abrupt termination of the planet's dark plain against such lava flows would seem to indicate that the plain itself has been partially covered with lava.

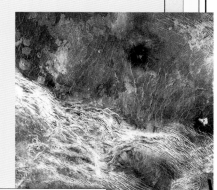

RIDGE BELTS

These three large craters, ranging from 23 to 31 miles (37–50 km) in diameter, are located in a region of fractured plains. They exhibit many features typical of meteor impacts, including bright ejecta, terraced walls, and central peaks. There are nearly 900 such craters on Venus—evidence of bombardment in its youth.

METEOR CRATERS

Small volcanic features, so-called pancake domes, are common on the surface of Venus. Most are about 7.5 miles (12 km) in diameter and have central pits like large volcanic cones. However, such domes are unique to the planet and were probably formed by single eruptions of extremely thick lava.

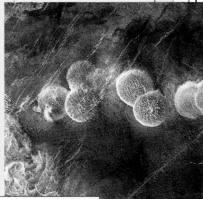

PANCAKE DOMES

MAGELLAN DATA

- Weight: 7,596 pounds (3,445 kg)
- Height: 21 feet (6.46 m)
- Diameter: 15 feet (4.61 m)
- Radar image resolution: 394–427 feet (120–300 m)
- Number of radar images per orbit: 1,852
- Area covered per orbit: 9,934 miles by 11–17 miles (15,996 km by 17–28 km)
- Surface measurements: up to 19 feet (30 m)

The Earth

See also:
- **The solar system** p. 32
- **Man on the Moon** p. 48

THE SEASONS

In the course of its 365.25-day orbit of the Sun, the Earth spins on an axis tilted at 23.5°. This accounts for the different seasons of the northern and southern hemispheres. During the four months of the northern summer, when the northern hemisphere is tilted toward the Sun, it receives more sunlight than the southern—and vice versa. By contrast, the equator experiences relatively little seasonal change, as do the extreme polar regions. The difference between the height of the Sun at midday in the northern or southern midwinters, and its height in midsummer, is a significant 47° (see below).

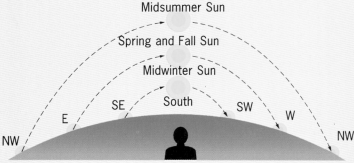

THE FIRST ASTRONAUTS in orbit, looking down at the Earth, were impressed by the sheer blueness of our planet. It was a reminder that seven-tenths of the surface is covered by water, and that it was water and oxygen, in association with volcanic action and a protective gaseous atmosphere, that made life possible. Such a combination of conditions was unique in the solar system, as the exploration of the other, lifeless planets has since confirmed. One Apollo astronaut memorably described the Earth as "a grand oasis in the vastness of space."

Right:
Earthrise over the lunar horizon, photographed by an Apollo 17 astronaut. A section through the planet (below) shows its thin silicate crust, rocky mantle, molten iron outer core, and solid iron inner core.

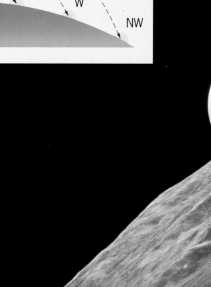

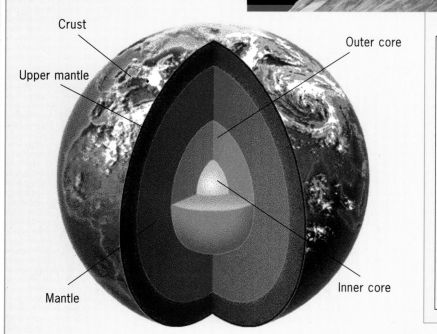

EARTH DATA

- Max. distance from Sun: 94 mil. mi (152 mil. km)
- Min. distance from Sun: 91 mil. mi (147 mil. km)
- Diameter: 7,921 miles (12,756 km)
- Orbits the Sun: once every 365.256 days
- Rotation period: 23 hrs, 56 mins, 4 secs
- Speed through space: 18.5 miles per sec (29.8 kps)
- Surface temperature: approx -128 °F to 120 °F (-89 °C to 49 °C)
- Atmosphere: 78 percent nitrogen, 21 percent oxygen, plus other gases, including carbon dioxide and argon

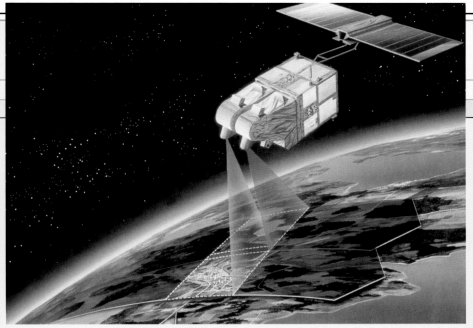

Above:
A "remote sensing" satellite that takes continuous views of the Earth in various wavelengths. Other satellites monitor weather systems such as hurricanes (right).

IMAGING THE EARTH

Satellites have revolutionized our understanding of the Earth. Not only do they help to forecast the weather, survey the atmosphere, and measure the effects of pollution, they enable us to monitor the resources of the planet. By observing the surface through infrared and other wavelengths, they can reveal mineral deposits and measure the health of soil. Satellite views are also used for mapmaking and provide valuable information for urban planners.

The European Community uses satellite imagery to monitor agriculture in member states. In this infrared image of a rural area in Great Britain, barley fields are registered as orange/brown, oilseed as pink, grassland as yellow/green, winter wheat as dark brown, and built-up areas as blue.

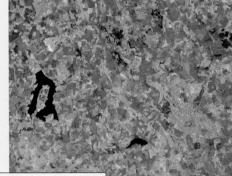

FIELD WATCH

A multispectral image of the Serbian capital, Belgrade. Such images can help urban planners to plot the route of a new highway or the location of a new manufacturing plant. The best sites for housing estates and even tourist attractions are also decided using remote sensing images. Pollution in rivers can also be monitored.

CITY VIEW

High-resolution satellite imagery can be combined with other data about specific areas of the Earth to create digital terrain maps. These three-dimensional views can be used by environmental planners to work out road routes or the best course for electricity pylons. They can also be used for military purposes.

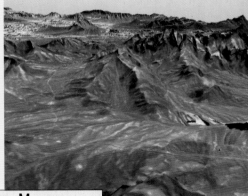

TERRAIN MAPS

The Moon

See also:
- **Exploring the Moon** *p. 46*
- **Man on the Moon** *p. 48*

LUNAR ORBITERS

The first images of the Moon's far side were taken in 1959 by the Russian *Luna 3* spacecraft. Over the following decade, five U.S. *Lunar Orbiters* (below) and a fleet of Russian *Luna* craft mapped both sides systematically. Launched between 1966 and 1967, the primary purpose of the U.S. *Lunar Orbiters* was to seek out suitable landing areas for manned Apollo craft. In the process, they returned many spectacular views of lunar craters and "seas." *Lunar Orbiter 4* was the first craft to orbit the Moon's poles.

FORMERLY A small planet that strayed too close to the Earth and was captured by its gravitational force, the Moon is our only natural satellite in space. Though visible at reliable intervals in its five principal phases—new, crescent, quarter, gibbous, and full—it shows only one face to us. When the side facing us is dark (a new Moon), the far side is fully illuminated by the Sun. The Moon is a completely dead world, heavily cratered by meteor impacts. In the great marias or "seas" there is no water, air, or weather, while gravity on the lunar surface is one-sixth of that on Earth. The satellite is 2,160 miles (3,476 km) in diameter and orbits the Earth at an average distance of 236,945 miles (381,308 km).

Below:
Lunar orbital sequence: the spacecraft approaches the Moon (1), fires a retro rocket (2), enters orbit and returns images to Earth (3). Another retro firing (4) places the orbiter in a lower orbit for close-up photography (5).

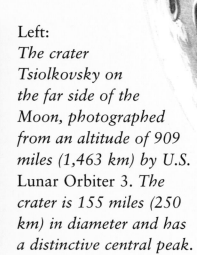

SEA OF SHOWERS

Archimed

Aristarchus

KEPLER

OCEAN OF STORMS

SEA C CLOU

Left:
The crater Tsiolkovsky on the far side of the Moon, photographed from an altitude of 909 miles (1,463 km) by U.S. Lunar Orbiter 3. The crater is 155 miles (250 km) in diameter and has a distinctive central peak.

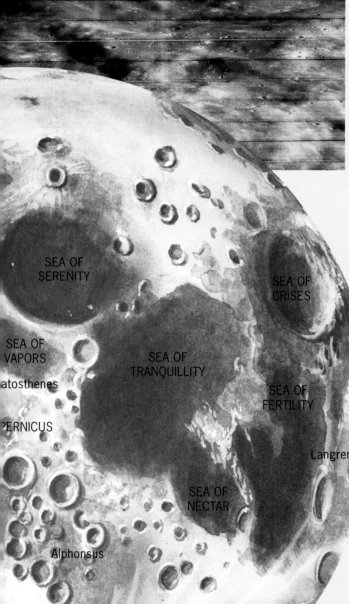

Earth and the
Moon compared

ORBITER IMAGES

Of all the pictures taken by the U.S. *Lunar Orbiters*, perhaps none
was more spectacular than that of the crater Copernicus in
November 1966 (left). Another picture revealed the crater Tycho.
One of the most prominent lunar craters, Tycho has a "ray" system
that can be seen from Earth with the naked eye. Further *Orbiter*
images include close-ups of the Hyginus crater, the volcanic
domelike Marius Hills, and the Hadley Rille, where *Apollo 15*
eventually landed.

Right:
*The 60-mile (97 km)
diameter crater
Copernicus as seen by an
Earth-based telescope.
Note the surrounding ray
system—debris scattered
by meteor impact.*

Below:
*The core of the Moon is
small in relation to its total
size, unlike the core of the
"solid" planets Mercury,
Venus, Mars, and Earth.*

Surface layer
of fine dust

Mantle

Small inner
core

Outer core

Below:
*The 53-mile (85
km) diameter
crater Tycho,
photographed by
Lunar Orbiter 5
from 137 miles
(220 km) up.*

SEA OF
SERENITY

SEA OF
CRISES

SEA OF
VAPORS

SEA OF
TRANQUILLITY

atosthenes

SEA OF
FERTILITY

PERNICUS

Langrenus

SEA OF
NECTAR

Alphonsus

Left:
*A map of the Moon
showing the large
marias, or "seas," and
several major craters.*

TYCHO

Exploring the Moon

See also:
- **Man on the Moon** *p. 48*
- **The Mars landings** *p. 52*

THE LUNA LANDERS

After the impact of the Soviet probe *Luna 2* (below) in 1959, the first soft landing on the Moon took place in 1966, when *Luna 9* touched down and sent back close-up views of the surface. After this, the Soviet Union's Luna program went from strength to strength. Another soft landing was made by *Luna 13* in December 1966, while part of *Luna 16* returned to Earth from the Moon's surface in 1970 carrying a small amount of lunar dust. Finally, *Luna 17* managed to deploy *Lunakhod*, the first unmanned roving vehicle on the Moon.

EXPLORATION OF THE MOON began in earnest as soon as it became technically possible to set a space probe down on the lunar surface. In September 1959 the first lunar probe, the Soviet Union's *Luna 2*, did not so much touch down on the surface as plunge into it at the edge of the Sea of Clouds. The succeeding unmanned lunar lander programs of both the Soviet Union and the U.S. were more sophisticated. They took close-up views of the surface and sent back data about the Moon's makeup and landing sites for projected manned craft.

Above:
A U.S. Atlas Agena B carries a Ranger lander to the Moon. Ranger 7 took close-ups of the lunar surface before impact (below).

Above:
The unmanned lunar roving vehicle Lunakhod, *and the tracks left by it (left). Its movements were remotely controlled from Earth. A second* Lunakhod *vehicle was deployed in 1973.*

UNMANNED MISSION DATA

Spacecraft	Launch date	Achievement
Luna 2	Sept. 12, 1959	First landing on the Moon
Luna 3	Oct. 4, 1959	First photos of far side
US *Ranger 7*	Jul. 28, 1964	First close-up photos
Luna 9	Jan. 31, 1966	First soft landing
Luna 10	Mar. 31, 1966	First lunar orbiter
Luna 16	Sept. 12, 1970	First sample return
Luna 17	Nov. 10, 1970	First roving vehicle

Cruise altitude

Positioning 30 minutes before touchdown

Main retro rockets start up to slow lander's descent

Main retro rocket burns out and is ejected

Engines shut off

Soft landing

THE SURVEYORS

Although the Soviet Union made the first soft landing on the Moon, the U.S. Surveyor program (right) was to overshadow the achievements of the *Lunas*. Seven *Surveyors* were launched, five of which made successful soft landings (see left) at sites being considered for future manned landings. The landing craft took close-up pictures of the surface and scooped up lunar soil using a robot arm. The last *Surveyor* landed near the crater Tycho in 1968.

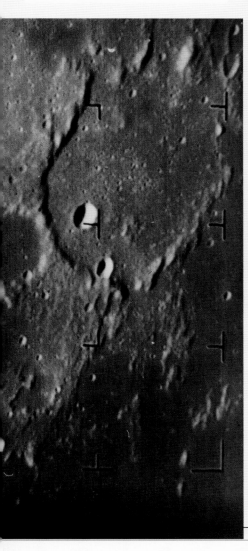

Above:
A "footprint" made by one of Surveyor 3's *landing pads when it bounced following touchdown on the Moon. Like* Surveyor 1 *in 1966 (right), it would have cast a large, spiderlike shadow over the lunar surface.*

Man on the Moon

See also:
- **The Moon** *p. 44*
- **Exploring the Moon** *p. 46*

LIFTING OFF

The Apollo astronauts were launched into space on the top of the mighty *Saturn 5* rocket. The booster was 364 feet (111 m) tall and had three stages. The first and second stages had five huge F-1 engines that generated a thrust of 7,501,410 pounds (3,402,000 kg). The third stage placed Apollo in Earth's orbit and later fired it toward the Moon.

Below:

Apollo landing profile. First the spacecraft entered lunar orbit and the Lunar Module separated from the Command Module. Then the Lunar Module landed on the Moon. Thereafter the LM ascent stage lifted off from the Moon to dock with the Command Module, which then returned to Earth.

Trans-Earth trajectory

LM ascent

Orbit and landing sequence

Launch and Earth orbit

Translunar trajectory

ON JULY 21, 1969, the astronaut Neil Armstrong stepped from the ladder of his *Apollo 11* Lunar Module onto the lunar surface. Few would disagree that the Moon landing was one of the most significant events in history—the culmination of years of scientific experiment, great human bravery, and billions of dollars' worth of investment. This technological feat was to be repeated five more times until the Apollo program came to an end in December 1972. Since then no one has returned to the Moon, and no lunar mission is planned in the near future.

Above:
The ascent stage of the Lunar Module returns from the Moon.

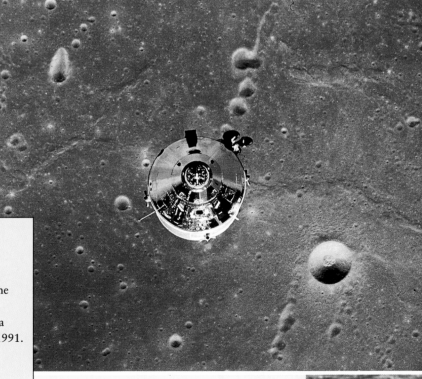

CREW PROFILES

- NEIL ARMSTRONG (*Apollo 11*) was a former *X-15* rocket-plane pilot and astronaut aboard *Gemini 8*.
- BUZZ ALDRIN (*Apollo 11*) made a two-hour space walk outside *Gemini 12*.
- PETE CONRAD (*Apollo 12*) flew two Gemini missions and commanded *Skylab*. He died in an accident in 1999.
- AL BEAN (*Apollo 12*) also flew a *Skylab* mission lasting 59 days in Earth's orbit.
- ALAN SHEPARD (*Apollo 14*) was America's first man in space. He died in 1998.
- EDGAR MITCHELL (*Apollo 14*) made only one spaceflight.

- DAVID SCOTT (*Apollo 15*) drove the first lunar rover.
- JAMES IRWIN (*Apollo 15*) became a Christian evangelist and died in 1991.
- JOHN YOUNG (*Apollo 16*) flew two Gemini, two Apollo, and two shuttle missions.
- CHARLIE DUKE (*Apollo 16*) also became a Christian evangelist.
- GENE CERNAN (*Apollo 17*) flew *Gemini 9* and an *Apollo 10* test flight.
- JACK SCHMITT (*Apollo 17*) was the first professional geologist on the Moon.

Above:
The Apollo 11 *Command and Service Module in orbit round the Moon—a photograph taken by the Lunar Module in orbit.*

The Apollo moon landing sites.

Left:
The Apollo 11 *crew: (from left) Neil Armstrong, Michael Collins, and Buzz Aldrin.*

MOON WALKS

The confidence of astronauts on the lunar surface grew as the Apollo program went on. The first moon walk (*Apollo 11*) lasted two hours and 21 minutes. The three *Apollo 17* moon walks, on the other hand, lasted a total of 22 hours and five minutes. Six crews stayed a total of 12 days and 11 hours at a variety of lunar sites.

Below:
The most famous Apollo 11 image: Buzz Aldrin standing at Tranquillity base on the Moon.

MISSION PROFILES

APOLLO 13

APOLLO 16

APOLLO 17

The *Apollo 13* crew (top) following their lucky escape in 1970 from an explosion in the Service Module: (from left) Jack Swigert, Fred Haise, and Jim Lovell. *Apollo 16* in 1972 was more successful, when astronaut Charlie Duke (middle) was able to make extensive use of the Lunar Roving Vehicle (LRV). On the last mission, *Apollo 17*, geologist Jack Schmitt (bottom) completed the program at Taurus Littrow.

Mars

See also:
- **The solar system** *p. 32*
- **The Mars landings** *p. 52*

MISSIONS TO MARS

Before July 1965, when the U.S. *Mariner 4* spacecraft (below) flew past Mars and took a series of grainy pictures, there were five failed Soviet attempts to send a spacecraft to the Red Planet. Six years later, *Mariner 9* became the first craft to orbit Mars. But the Soviet Union's Mars program continued to be unsuccessful. Thirteen failures were recorded before its first relatively successful *Mars 5* flight in 1973. Subsequent Russian attempts have also ended in failure.

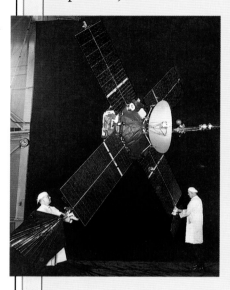

K**NOWN** AS the Red Planet because of its distinctive ruddy color in the night sky, Mars has always been the subject of great speculation. Early telescopic observations indicated that parts of Mars changed shape and surface texture, and strange linear features led to suggestions that the planet was inhabited. The notion of "life on Mars" persists, though the visits of numerous spacecraft have finally discredited the idea of Martian aliens. More Mars explorations are planned, possibly leading to the return of Martian soil and rock samples to Earth.

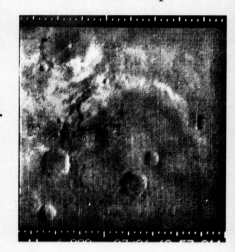

Above:
The first relatively clear image of Mars, taken by Mariner 4 in July 1965.

Above:
Phobos, *a Russian craft launched in 1988 to explore Mars and its two moons, Phobos and Deimos. The craft developed faults as it approached the planet.*

MARS DATA

- Max. distance from Sun: 155 mil. mi (249 mil. km)
- Min. distance from Sun: 128 mil. mi (207 mil. km)
- Diameter: 4,215 miles (6,787 km)
- Orbits the Sun: once every 687 days
- Rotation period: 24 hrs 37 mins
- Speed through space: 15 miles per second (24.1 kps)
- Mean surface temperature: -9.4° F (-23° C)
- Atmosphere: carbon dioxide
- Phobos dimensions: 17 by 12 miles (27 by 19 km)
- Deimos dimensions: 6 by 7 miles (10 by 12 km)

Earth and Mars
compared

Right:
*The Martian surface,
a world of volcanoes
and plains riven by
huge canyons.*

MARTIAN LANDSCAPE

SURFACE FEATURES

A variety of Martian features have been closely
studied by visiting spacecraft. *Mariner 4* first
spotted craters, while *Mariner 9* discovered
the huge Olympus Mons volcano. Two *Viking*
orbiters conducted a systematic survey of the
planet, including the polar caps. More
recently, the *Mars Global Surveyor* (in orbit
since 1997) has returned spectacular close-ups
of the so-called "Mars face," proving that it is
simply a mesa (table-shaped hill) with
suggestive shadows.

Far taller than Mt. Everest,
the volcano Olympus Mons
towers 13.5 miles (22 km)
into the Martian sky and is
about 342 miles (550 km) in
diameter. It has a caldera
(volcanic crater) in its
summit, and is covered by
thousands of individual
lava flows extending
hundreds of miles (km) over
the surrounding terrain.

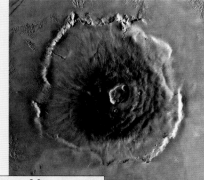

OLYMPUS MONS

A *Mars Global Surveyor*
close-up of the Martian
north pole, where ice cliffs
are up to 2 miles (3.5 km)
thick and have spiral patterns made by wind ero-
sion. Shrinking and expanding according to seasonal
variations in temperature,
the planet's polar caps are
composed largely of ice
and frozen carbon dioxide.

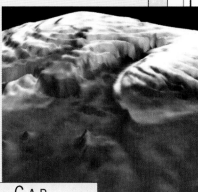

POLAR CAP

Below:
*The Valles Marineris canyon. It stretches
a quarter of the way around Mars—
equivalent to the width of the U.S.—and is
thought to have once contained water.*

Phobos and Deimos are
thought to have been
asteroids captured by
Martian gravity. Phobos is
in an unusual orbit. Moving
from east to west, it
crosses the Martian sky in
only four and a half hours,
and appears again 11
hours later. Deimos remains
in the Martian sky for two
and a half Martian days.

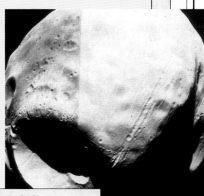

PHOBOS

The Mars landings

See also:
- **Mars** p. 50
- **Mapping Venus** p. 40

EARLY EXPLORATION

In the early days of Mars exploration, touchdown on the planet proved an elusive goal. Starting in 1969, seven Russian attempts failed. One Russian craft, *Mars 6*, appeared to have landed safely in 1974—only to be covered by its parachute before it could send back data. The U.S. Viking program (left) met with more immediate success, the two landers touching down in July and September 1976. Both had a robot arm to collect Martian soil for analysis.

A PART FROM seeing the first men walking on the Moon, nothing in space exploration has captured the public imagination more than the Mars landings. When the three U.S. spacecraft, *Vikings 1* and *2* and the *Mars Pathfinder,* landed on the Red Planet, contrary to "life on Mars" theories, they found a sterile world. Yet more exploration did show that parts of Mars were once covered in water. This fired up enthusiasm for further expeditions and, with the constant improvement in our ability and technology, the future of Mars exploration looks bright.

Above:
The robot arm of the Viking *lander. No sign of life was found in the soil samples taken.*

Above:
The Sojourner, *the* Mars Pathfinder's *small planetary rover.*

Left:
Ares Vallis, the Mars Pathfinder's *landing site. The spacecraft's camera took numerous stunning panoramas of the Red Planet's surface (below).*

FLIGHT LOG

- *Viking 1*: launched Aug. 20, 1975; landed July 20, 1976; last transmission Nov. 1982
- *Viking 2*: launched Sept. 9, 1975; landed Sept. 3, 1976; last transmission April 1980
- *Mars Pathfinder*: launched Dec. 4, 1996; landed July 4, 1997; last transmission Nov. 1997
- *Mars Climate Orbiter*: launched December 11, 1998; lost on arrival September 23, 1999
- *Mars Polar Lander*: launched January 3, 1999; lost on arrival December 3, 1999

STUDYING THE SURFACE

The *Viking 1* and *2* spacecraft were well-equipped to study the Martian environment. They recorded that temperatures ranged from about -123 °F (-86 °C) before dawn to -27 °F (-33 °C) in the mid-afternoon, and that gusts of wind could reach 31 miles per hour (50 kph). They took pictures showing a pink sky, caused by reddish dust scattering sunlight, and spectacular sunsets, in which there were a number of haloes. Carbon dioxide ice frost was also photographed on rocks.

Top, above and left:
Three Viking *pictures of the rocky Martian surface. One of the spacecraft's landing pads is visible in the image above.*

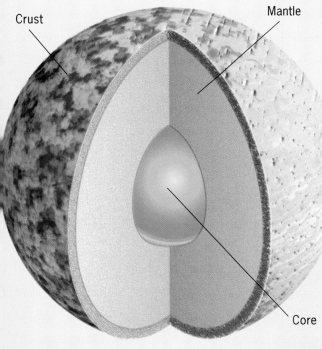

Crust

Mantle

Core

Left:
A section of the Red Planet—about one third the size of Earth.

Below:
An impression of the Mars Polar Lander *following touchdown in December 1999.*

VISITORS TO MARS

The exploration of the planet Mars has not been totally successful. The *Mars Climate Orbiter* was lost on arrival on September 23, 1999, a disaster that was followed three months later by the loss of the *Mars Polar Lander*, also known as *Deep Space 2*. Current projects include *Mars Global Surveyor* and *2001 Mars Odyssey*, both launched in 2001 and now on their way to Mars. Several more probes are planned as far ahead as 2005. The next launch, of *Mars Express*, is scheduled for 2003.

Asteroids and meteors

See also:
- **Space observation** p. 10
- **Comets** p. 72

ASTEROID DIMENSIONS

With a diameter of between 559 and 621 miles (900–1000 km), Ceres (below) is the largest known asteroid. Deeply cratered by the impacts of other rocks, it orbits the Sun every 4.6 years and was first observed on January 1, 1801. The next largest asteroid, with a diameter of 373 miles (600 km), is Pallas. Other large asteroids are Vesta (diameter 329 miles/530 km) and Hygela (diameter 280 miles/450 km). But these monsters are exceptions to the rule—the majority of asteroids are less than 0.6 miles (1 km) across.

ASTEROIDS AND METEORS are matter left over from the formation of the solar system. But while asteroids are sizable chunks of rock—a few are immense—meteors are usually no more than large stones or dust particles, thousands of which burn up in the Earth's atmosphere every day. Those meteors that do survive the passage are known as meteorites, and their impacts can be violent. But any asteroid collision with Earth would be far more catastrophic, so they are carefully monitored. There are three main groups of asteroids. The near-Earth asteroids intersect the Earth's orbit when they travel close to the Sun, while the Trojans orbit in two clusters on either side of Jupiter. The so-called "main belt" asteroids orbit the Sun between Mars and Jupiter.

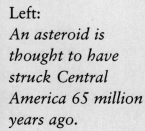

Left:
An asteroid is thought to have struck Central America 65 million years ago.

Right:
The main belt asteroids orbit between the orbits of Mars and Jupiter, while the Trojans are in the same orbit as Jupiter.

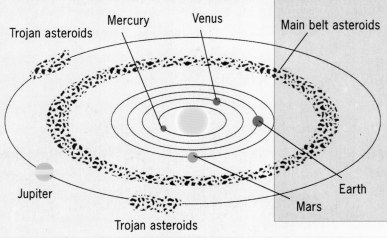

Trojan asteroids • Mercury • Venus • Main belt asteroids • Jupiter • Earth • Mars • Trojan asteroids

ASTEROID FACTS

- Estimated number in solar system: 50,000
- Number between Mars and Jupiter: 2,500
- First to be explored by spacecraft: Gaspra
- Most recent impact on Earth: 1908, Siberia; however, this might have been a comet
- Weight of largest found on Earth: 60 tons
- Brightest in night sky: Vesta

Earth and Ceres compared

METEOR SHOWERS

A total of 22 meteor showers occur in the Earth's atmosphere during nine months of the year. The meteor trails can be traced back to one point in the sky, which is known as the radiant. The meteor shower is named after the constellation in which the radiant appears. So the Leonids, which appear in November (right), take their name from the constellation Leo. Other well-known showers are the Perseids and Orionids.

Above left:
The asteroid Gaspra, photographed by the spacecraft Galileo *in 1991.* Galileo *also flew past 32-mile (52 km) wide Ida (above right with Gaspra to scale) in 1993, and inspected its surface.*

Left:
The Tswaing/Soutpan Meteor Crater near Pretoria, South Africa. The meteor may have originated from the main belt of meteors intersecting our orbit (below).

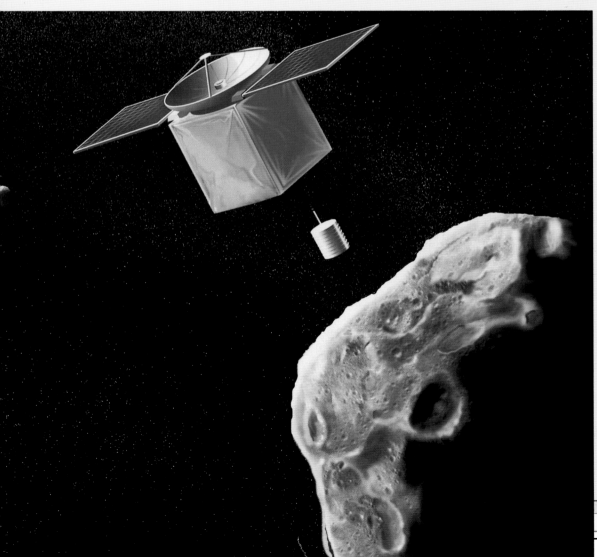

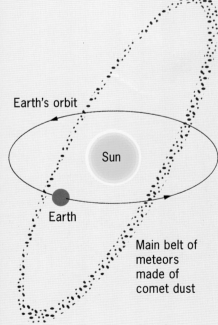

Earth's orbit

Sun

Earth

Main belt of meteors made of comet dust

Left:
Scheduled for launch in 2002, the Japanese Muses-C mission will make a landing on near-Earth asteroid 1998 SF36.

Jupiter

See also:
- **The solar system** *p. 32*
- **Jupiter's moons** *p. 58*

Below:
Jupiter's banded appearance is created by areas of rising and falling gas.

J UPITER IS THE biggest planet in the solar system. Although just half its mass is greater than that of all the other planets combined, with compressed hydrogen and helium in place of a solid planetary surface, this giant's density is lower than the Earth's. Jupiter's atmosphere is made up of multicolored gases whipped along by winds of over 400 miles per hour (650 kph). It has four large moons, first observed by Galileo— though at least 12 other smaller moons and a small ring system have since been discovered. Five spacecraft have explored the planet. One of these, *Galileo*, went into orbit in 1995 and deployed a capsule that plunged through Jupiter's atmosphere.

Left:
Pioneer 10, *the first spacecraft to fly past Jupiter in 1973 (see below left). For its mission, Galileo was deployed in Earth's orbit by the space shuttle (right). Jupiter has been visited by Pioneer 11 and Voyagers 1 and 2.*

Detector panels

Magnetic field sensor

Atomic power source

PIONEER 10 MISSION

Launch: March 3, 1972

Asteroids

Earth

Mars orbit

Pioneer 10 reaches Jupiter December 5, 1973

JUPITER DATA

- Max. distance from Sun: 506.5 mil. mi (815.7 mil. km)
- Min. distance from Sun: 460 mil. mi (740.9 mil. km)
- Equatorial diameter: 88,679 miles (142,800 km)
- Polar diameter: 83,214 miles (134,000 km)
- Rotation period: 10 hours, the fastest of the planets
- Orbits the Sun: once every 11.86 years
- Speed through space: 8.1 miles per second (13.06 kps)

Earth and Jupiter compared

THE GREAT RED SPOT

Jupiter's Great Red Spot (below) is an atmospheric storm formed by gases swirling at speeds of up to 224 miles per hour (360 kph). Located in the planet's southern hemisphere, it measures 20,000 by 7,450 miles (32,000 by 12,000 km) and towers 5 miles (8 km) above the surrounding clouds. First observed by the astronomer Robert Hooke in 1664, the storm varies in intensity and coloration through the year. The red color of the Spot indicates an extremely large quantity of phosphorus.

Above:
The Galileo *probe, which entered Jupiter's atmosphere at a speed of 29 miles per second (47 kps) and sent back data from the hydrogen interior (right).*

- Rocky core
- Metallic hydrogen
- Liquid hydrogen
- Hydrogen gas

Below:
This image from the Galileo *orbiter reveals the beautiful mosaic of colors in Jupiter's atmosphere. Nine distinct bands of clouds circulate the planet.*

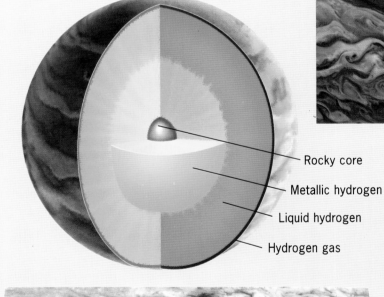

Above:
An impression of Jupiter's cloud tops as they might look to a spacecraft speeding over them.

Jupiter's moons

See also:
- **Jupiter** *p. 56*
- **The solar system** *p. 32*

VOLCANIC IO

Jupiter's largest moon, Io, veers dangerously close to Jupiter in the course of its elliptical orbit. Its surface is affected by the pull of the planet's gravity —it actually "bulges"—and the matter at its core is heated to tremendous temperatures. Then, as the moon moves away, its core cools and its surface contracts. Such violent changes have given birth to numerous volcanoes on Io, whose constant eruptions send sulfur flying into space and cover the surface with molten lava. The largest volcano is known as Pele.

WHEN GALILEO first observed Jupiter, he also saw four small points of light nearby that constantly shifted position. These were the planet's four largest moons—Europa, Io, Callisto, and Ganymede. Each of these moons has a distinctive character. While Europa is covered by an ice pack, Io is an inferno of volcanoes that spew sulfur 186 miles (300 km) into space. Ganymede, on the other hand, has a cracked surface, and Callisto has a deeply pitted one. Before spacecraft explored the Jupiter region, it was thought there were also another nine smaller moons. In fact there may be 28 in orbit round the planet.

Above:
A close-up image of a crater chain on the moon Callisto.

EUROPA

Above:

Thirteen of Jupiter's sixteen moons. Five were discovered visually while the others were recorded photographically; the smaller moons have all recently been pictured by Voyager.

THE MOONS OF JUPITER

NAME	DISTANCE FROM JUPITER	DIAMETER
Metis	79,460 mi/127,873 km	25 mi/40 km
Adrastea	80,096 mi/128,896 km	12 mi/20 km
Amalthea	112,590 mi/181,188 km	123 mi/198 km
Thebe	137,800 mi/221,757 km	62 mi/100 km
Io	261,814 mi/421,600 km	2,272mi/3,659km
Europa	416,629 mi/670,900 km	1,894mi/3,050km
Ganymede	664,470 mi/1.07 mil. km	3,273mi/5,270km
Callisto	1.2 mil. mi/1.88 mi. km	3,106mi/5,000km
Leda	6.9 mil. mi/11.1 mi km	5 mi/8 km
Himalia	7.1 mil. mi/11.47 mil. km	106 mi/170 km
Lysithea	7.27 mil. mi/11.71 mil. km	12 mi/19 km
Elara	7.3 mil. mi/11.74 mil. km	50 mi/80 km
Ananke	12.8 mil. mi/20.7 mil. km	11 mi/17 km
Carme	13.9 mil. mi/22.35 mil. km	15 mi/24 km
Pasiphae	14.5 mil. mi/23.3 mil. km	17 mi/27 km
Sinope	14.7 mil. mi/23.7 mil. km	13 mi/21 km

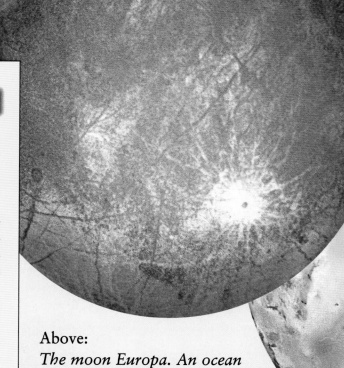

Above:
The moon Europa. An ocean is thought to lie below its covering ice pack.

CRACKS IN EUROPA'S ICE

GANYMEDE

Like Io, Europa orbits close enough to Jupiter to be affected by the planet's gravitational force. This has created tidal movements in the moon's ocean, which in turn have sent cracks often hundreds of miles (kilometers) long through the covering ice pack. The ocean's water may also be heated slightly in the process. This has led some scientists to suggest it might harbor primitive living cells.

Left:
An impression of flying low over the moon Ganymede.

Below:
The sulfur surface of the moon Io, whose many volcanoes erupt continuously.

Above and left:
Ganymede, the largest moon in the solar system, and Callisto, its surface pitted with meteor impact craters.

CALLISTO

IO

Right:
An artist's impression of a possible future craft built to plunge through Europa's ice pack and go in search of life-forms in the moon's ocean.

Saturn

See also:
- **The solar system** p. 32
- **Saturn's moons** p. 62

EXPLORING SATURN

The first spacecraft to fly past Saturn, at a distance of 13,000 miles (21,000 km), was *Pioneer 11* (September 1, 1979). Its successor *Voyager 1* flew past from 77,000 miles (124,000 km) on November 12, 1980, while *Voyager 2* flew past from 62,720 miles (101,000 km) on August 26, 1981. But *Cassini* will be the first spacecraft to orbit Saturn and explore many of its moons (below). At one point, *Cassini*'s probe *Huygens* will be deployed to land on the largest moon, Titan.

S ATURN IS the second largest planet in the solar system. It is also one of the most beautiful objects in space. Though its brightly colored rings have fascinated astronomers for over 300 years, it was not until the spacecraft *Pioneer 11* and *Voyagers 1* and 2 flew past Saturn (1979–81) that their secrets were revealed. At the same time, wonderful close-up views of the planet itself showed an atmosphere consisting mainly of hydrogen. Like Jupiter, Saturn rotates so fast (once every 10 hours 40 minutes) that it bulges slightly at its equator. In 1997 a new spacecraft, *Cassini*, was launched to investigate the planet again. Packed with research instruments, it is the size of a bus. It is due to arrive in 2004 and will go into orbit, surveying the planet, its rings, moons, and magnetic fields for about three years.

Above:
Cassini is launched on a Titan 4B *booster from Cape Canaveral.*

Below:
Beneath its hydrogen clouds, Saturn is thought to have a layer of liquid hydrogen and a layer of metallic hydrogen. The core of the planet is believed to be solid rock.

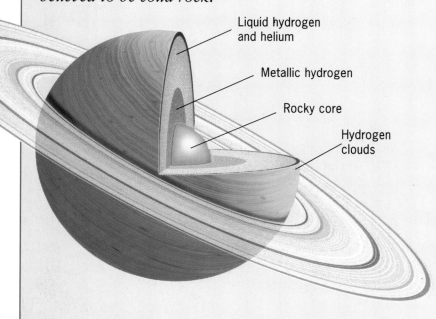

Liquid hydrogen and helium

Metallic hydrogen

Rocky core

Hydrogen clouds

SATURN DATA

- First telescopic observation: Galileo (1610)
- First detailed telescopic observation of ring system: Christian Huygens (1659)
- Max. distance from Sun: 936 mil. mi (1,507 mil. km)
- Min. distance from Sun: 837 mil. mi (1,347 million km)
- Equatorial diameter: 74,893 miles (120,600 km)
- Polar diameter: 66,880 mi (107,700 km)
- Orbits the Sun: once every 29.46 years
- Rotation period: 10 hrs 40 mins
- Speed through space: 6 miles per second (9.6 kps)
- Temp. of surface atmosphere: -292 °F (-180 °C)

Right:
A Voyager close-up of the upper atmosphere of Saturn, showing stormy areas and white clouds.

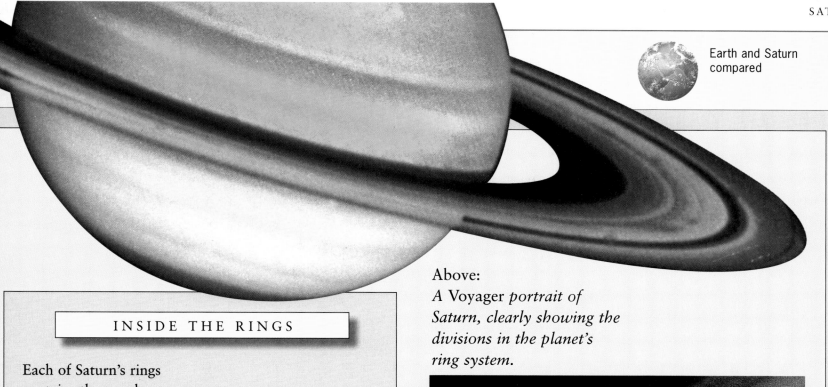

Earth and Saturn
compared

INSIDE THE RINGS

Each of Saturn's rings
contains thousands
of individual
"ringlets" made
up of small pieces
of ice and rock
(right). Such ring material
may range in size from 33-foot (10 m)
wide "icebergs" to specks of dust, and is
held together in orbit by both the gravity of the
planet and a series of small "shepherd moons." The
ring system itself is 168,900 miles (272,000 km) wide
but a mere 98 feet (30 m) high.

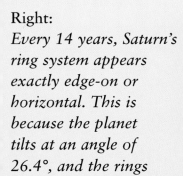

Above:
A Voyager *portrait of
Saturn, clearly showing the
divisions in the planet's
ring system.*

Above:
This Voyager *close-up reveals a number of
Saturn's moons.* Voyager *and* Pioneer *mission
telescopic observations suggest there may be as
many as 24 moons.*

Right:
*Every 14 years, Saturn's
ring system appears
exactly edge-on or
horizontal. This is
because the planet
tilts at an angle of
26.4°, and the rings
tilt with it.*

Saturn's moves

See also:
- **Jupiter's moons** *p. 58*
- **Saturn** *p. 60*

EXPLORING TITAN

When the spacecraft *Cassini* arrives in Saturn's orbit in 2004, it will begin to take close-up images of the planet's ring system and moons. Near Titan it will also deploy the probe *Huygens* (left)—named after the astronomer who discovered the Moon in 1655—that will land and take measurements.

Below:
A montage of Voyager *images of Saturn and six of its moons: (left to right) Rhea, Enceladus, Dione, Tethys, Mimas, and Titan.*

SATURN HAS AT LEAST 24 moons, the largest of which is Titan. Like the Earth, this 3,192-mile (5,140 km) diameter moon has a dense atmosphere of mainly nitrogen. But the similarity ends there. With a surface temperature of -292 °F (-180 °C), atmospheric pressure twice that of our planet, and perhaps great lakes of liquid methane, Titan would seem utterly hostile to life. Of the other moons, Prometheus, Pandora, and Atlas are perhaps the most intriguing, having a role out of proportion to their small size. It is thought the gravity of these "shepherd moons" helps to keep the rings of Saturn in place.

Above:
A Hubble Space Telescope image of Titan.

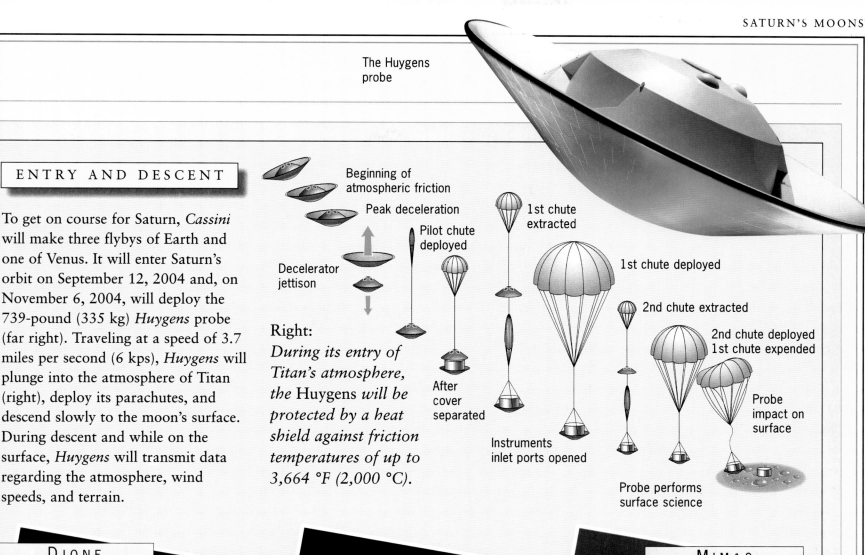

The Huygens
probe

ENTRY AND DESCENT

To get on course for Saturn, *Cassini* will make three flybys of Earth and one of Venus. It will enter Saturn's orbit on September 12, 2004 and, on November 6, 2004, will deploy the 739-pound (335 kg) *Huygens* probe (far right). Traveling at a speed of 3.7 miles per second (6 kps), *Huygens* will plunge into the atmosphere of Titan (right), deploy its parachutes, and descend slowly to the moon's surface. During descent and while on the surface, *Huygens* will transmit data regarding the atmosphere, wind speeds, and terrain.

Beginning of atmospheric friction

Peak deceleration

Decelerator jettison

Pilot chute deployed

1st chute extracted

1st chute deployed

2nd chute extracted

2nd chute deployed
1st chute expended

After cover separated

Instruments inlet ports opened

Probe impact on surface

Probe performs surface science

Right:
During its entry of Titan's atmosphere, the Huygens *will be protected by a heat shield against friction temperatures of up to 3,664 °F (2,000 °C).*

DIONE

ENCELADUS

MIMAS

Above:
Three of Saturn's moons: Dione (which may have ice volcanoes), Enceladus (the most reflective), and Mimas (the most heavily cratered).

Right:
An impression of the Huygens *probe landing on Titan's swampy surface of liquid gases, under an orange-red smog that rains liquid methane.*

Uranus

See also:
- **The stargazers** *p. 8*
- **The solar system** *p. 32*

THE DISCOVERY

Uranus was discovered by the astronomer William Herschel in 1781. Using a telescope in his back garden in Bath, England, Herschel compared the position of this tiny greenish disc in the night sky against a star map, and noticed during the following days that it appeared to move in relation to the stars. From this he concluded that he had discovered "either a nebulous star or perhaps a comet," and it was only some time later that he realized this was in fact the seventh planet. Appropriately, Uranus was named after the ancient Greek god of the sky.

Right:
Herschel's 40-foot (12 m) telescope, which he used to discover the two moons Oberon and Titania.

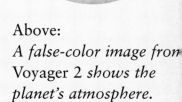

P LANETS CLOSE to the Sun, and so visible to the naked eye, have been known since ancient times. Uranus, one of the more remote planets orbiting the Sun, was discovered by telescope in 1781. Some 32,168 miles (51,800 km) in diameter, Uranus is a green-colored gas planet formed mainly of hydrogen and helium, with traces of methane. The planet is remarkable for a number of other reasons. It has a system of rings, five main moons— Miranda, Ariel, Umbriel, Titania, and Oberon— and rotates on its side in a counterclockwise direction. In 1986 Uranus was visited by the legendary deep-space probe, *Voyager 2.*

Above:
A false-color image from Voyager 2 *shows the planet's atmosphere.*

Right:
Voyager 2 *took close-up images of the thin ring system, which appears to contain particles less than 3 feet (1 m) in diameter.*

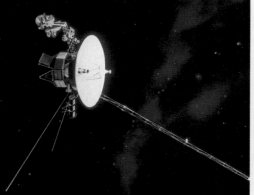

Above:
Voyager 2 *passed Uranus on January 26, 1986 (right), using the planet's gravity to set a course for Neptune.*

+8 hr

0 hr

Miranda

Oberon

Umbriel

Triton

Ariel

-8 hr

URANUS DATA

- Max. dist. from Sun: 1,865 mil. mi (3,004 mkm)
- Min. dist. from Sun: 1,782 mil. mi (2,869 mkm)
- Diameter: 32,168 miles (51,800 km)
- Orbits the Sun: once every 84.01 years
- Rotation period: 17 hrs 14 min
- Speed through space: 4.2 mi. per sec (6.8 kps)
- Angle of tilt: 98°

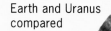
Earth and Uranus
compared

THE TILTED PLANET

The poles of Uranus are almost at right angles to the Sun, which
the planet orbits once every 84 years. This strange axis of rotation
means that the poles alternate between 42 years of daylight and
42 years of night. The reasons for the planet's tilt and its
counterclockwise rotation are hard to fathom. Possibly they came
about when another large celestial body cannoned into it during
the early history of the solar system.

Left:
*The strange surface of the
moon Miranda, made up
of ravines, mountains,
and craters.*

Rocky core

Ice

Molecular
hydrogen

Axial tilt
98°

Above:
*An illustration of
the planet's unusual
tilt and rotation.*

Right:
*Two moons of
Uranus—Ariel,
which has been
resurfaced in places
by eruptions of
lava, and Titania,
which is heavily
pocked with
impact craters.*

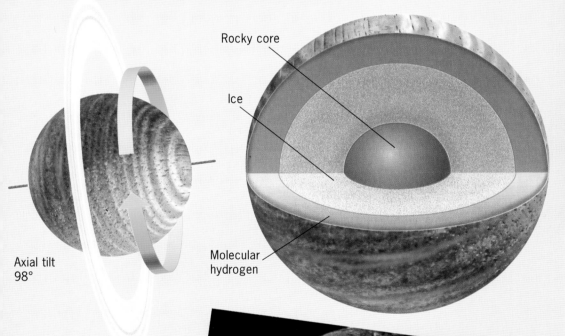

Above:
*A final view of Uranus as
Voyager 2 speeds away
toward Neptune.*

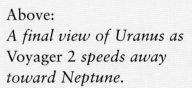

ARIEL

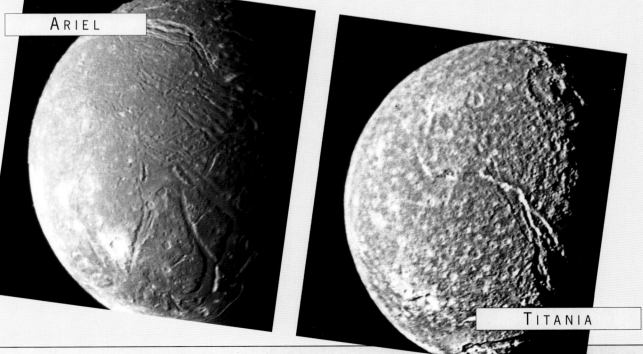

TITANIA

Neptune

See also:
- **The solar system** p. 32
- **Uranus** p. 64

NEPTUNE is a relatively recent discovery. Though some early 19th-century astronomers had made mathematical predictions of its presence near Uranus, it was not until 1846 that the German astronomer Johann Galle was actually able to observe it. Named after the Roman sea god, this icy-blue planet is one of the most dynamic in the solar system, rotating once every 19 hours. Its atmosphere of methane, helium, and hydrogen spins even faster, and is crossed by "scooter clouds" traveling at over 1,242 miles per hour (2,000 kps). On August 24, 1989, the spacecraft *Voyager 2* flew past the planet. It discovered a small ring system and six moons in addition to the two already known.

Above:
Whitish "scooter clouds," propelled by violent winds in Neptune's atmosphere.

Left:
The coldest known place in the solar system: Triton, where the temperature is -389 °F (-235 °C).

Below:
The flight path of Voyager 2, which flew past Neptune and its moons at a speed of 16.7 miles per second (27 kps).

Above:
The crescent of Neptune—a farewell shot by Voyager 2 following its flyby of the planet.

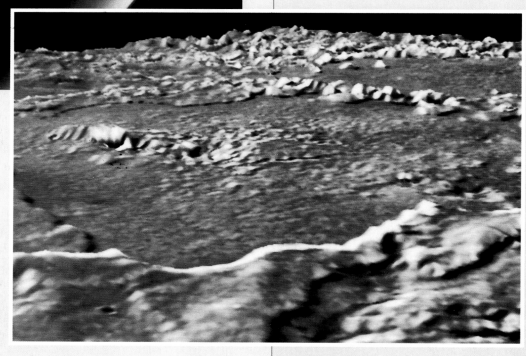

NEPTUNE DATA

- Max. distance from Sun: 2,817 mil. mi (4,537 mil. km)
- Min. distance from Sun: 2,767 mil. mi (4,456 mil. km)
- Orbits the Sun: once every 164.8 years
- Diameter: 30,740 miles (49,500 km)
- Speed through space: 3.4 miles per sec (5.43 kps)

23,600 mile (38,000 km) flyby of Triton

Closest point to Neptune 2,980 miles (4,800 km)

Neptune

Triton

Voyager 2 flight path

Earth and Neptune
compared

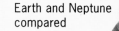

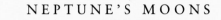

GREAT DARK SPOT

CIRRUS CLOUDS

Left:
*Neptune's
Great Dark
Spot, an Earth-
sized storm in
the planet's
atmosphere.*

Above:
*Cirrus clouds of
frozen methane,
casting shadows
on the atmo-
sphere 31 miles
(50 km) below.*

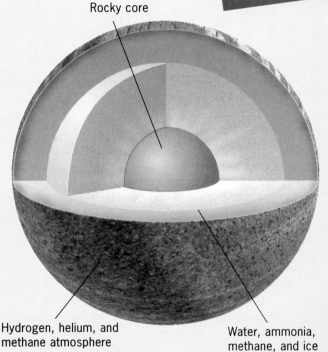

Rocky core

Right:
*It is thought
that an ocean of
water, ammonia,
and methane
lies beneath
Neptune's
atmosphere.*

Hydrogen, helium, and
methane atmosphere

Water, ammonia,
methane, and ice

NEPTUNE'S MOONS

Two of Neptune's moons, Triton and Nereid, were discovered by telescope. A world of liquid nitrogen
swamps and geysers, Triton (below, with Neptune in
the distance) has a diameter of 1,864 miles (3,000 km)
and is about six times larger than Nereid. Both moons
have remarkable orbits. While Triton's is circular and
moves east to west, Nereid is in a very deep orbit in
the opposite direction. They may have been thrown
out of their original paths by Pluto and its companion,
Charon, straying too close. Neptune's six other moons
were found by *Voyager 2*.

Left:
*The moon Triton, an elemental chaos
of cliffs, craters, mountains, glaciers, and
spouting liquid nitrogen geysers.*

Pluto

See also:
- **The solar system** p. 32
- **Neptune** p. 66

RECENT DISCOVERY

Pluto was the last planet to be discovered. On February 18, 1930, the American astronomer Clyde Tombaugh was comparing photographs of stars in the constellation Gemini when he noticed that one of the "stars" had shifted position (below). He concluded that it could only be an orbiting planet—confirming the suspicions of earlier astronomers that one might exist beyond Neptune. Pluto has a moon, Charon, which was discovered in 1978. It is thought both were once moons of the planet Neptune, because they occasionally come within Neptune's orbit.

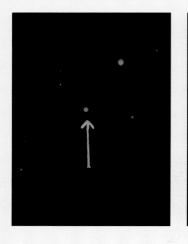

PLUTO IS THE outermost planet in the solar system. It is so far from the Sun—a mean distance of 3,664 million miles (5,900 million km)—that it takes 90,465 days to make one orbit. In fact it will not reach the orbital point at which it was discovered in 1930 until the year 2177. A small planet, Pluto is thought to consist of mainly rock and ice. The *Hubble Space Telescope* has taken the clearest images of the surface, but even these show only general light and dark areas. When it was first discovered (see left), the planet's mass and density were unknown. The uncovering of Charon by Jim Christy in 1978 enabled scientists to calculate the mass of Pluto by studying the orbit of its moon. Believed to be the only natural satellite, Charon orbits Pluto on a circular path in 6.4 days, and its diameter is thought to be half that of Pluto; about 714 miles (1,150 km). The moon is named after the mythological ferryman of the river Styx in the Greek underworld, Hades.

Right:
A Hubble image of Pluto and Charon. It is thought that the ice on the planet's surface (below) also contains frozen methane and ammonia.

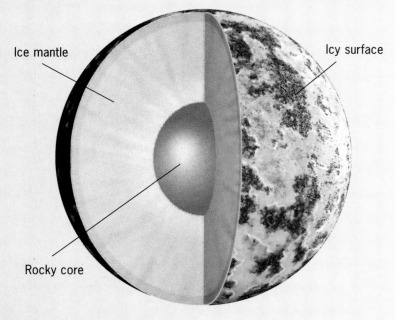

Ice mantle

Icy surface

Rocky core

PLUTO DATA

- Max. distance from Sun: 4,580 mil. mi (7,375 mil. km)
- Min. distance from Sun: 2,748 mil. mi (4,425 mil. km)
- Diameter: 1,428 miles (2,300 km)
- Orbits the Sun: once every 247.7 years
- Rotation period: 6 days, 9 hrs, and 17 mins
- Speed through space: 2.9 miles per second (4.7 kps)
- Surface temperature: -382 °F (-230 °C)
- Diameter of Charon: 714 miles (1,150 km)
- Distance Pluto to Charon: 12,233 mi (19,700 km)

Earth and Pluto
compared

Left:
*An artist's impression of the
planet Pluto and its moon,
Charon, in orbit around
the Sun.*

Charon

Path of
Pluto's
solar orbit

Pluto

THE KUIPER BELT

Beyond the orbit of Pluto lies a mysterious region of space
that is slowly starting to reveal its secrets. Here there lies a
belt of small rocky objects slowly moving around the Sun.
This is known as the Kuiper Belt, after the Dutch-American
astronomer who discovered it in 1951. Each object is a few
miles (km) across and they are thought to be the
frozen leftovers from ancient planet formation. Several hundred have been seen
through infrared telescopes.

Right:
*To an observer on Pluto,
the moon Charon would
look stationary. This is
because Charon's orbital
time—6.3 days—is exactly
the same as Pluto's
rotation time.*

Left:
*An impression of
Pluto's forbidding
surface.*

BLURRED VISION

This map of 85 percent of Pluto's surface (above) is derived
from *Hubble Space Telescope* images. It shows bright polar
areas, a dark equatorial belt, and the complex distribution of
frosts that migrate across the planet's surface according to
various seasonal cycles. In reality, Pluto's surface is probably
even more varied than is shown here—the *Hubble*'s resolution
tends to blur edges and blend small and large features.

Below:
*Pluto's orbit of the
Sun—elliptical as
well as sharply
inclined.*

Neptune

Pluto's orbit is
inclined and at times
comes within
Neptune's orbit

Pluto

Comets

See also:
- **Asteroids and meteors** p. 54
- **The Hubble Space Telescope** p. 76

COLLISION WITH JUPITER

In 1992, the comet Shoemaker Levy 9 flew close to the giant planet Jupiter and disintegrated. Fragments of it were spotted by the *Hubble Space Telescope* (bottom). It was predicted that these pieces would hurtle toward Jupiter again in July 1994 and actually collide with the planet. The *Hubble* was ready to capture this historic event. As the comet fragments plunged through Jupiter's clouds, shock waves more powerful than a series of nuclear blasts created huge ripples and dark spots on the planet's surface (left).

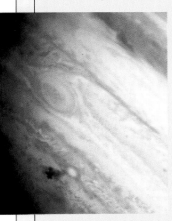

THROUGHOUT HISTORY men have seen what they took to be strange stars blazing across the night sky. In fact they were comets, which are miles-wide bodies of rock, ice, and dust left over from the formation of the solar system. Comets are deep-space wanderers, visible to us only when they near the Sun. Then the Sun boils the comet's ice, creating a cloud of shining dust and gas. A "tail" of such material then streams away from the comet. Comets are formed far out on the very edge of the solar system, where an enormous cloud, known as the Oort Cloud, can be found. This consists of billions of frozen comets, which are sometimes pushed out of their orbit by the gravity of the stars surrounding them, causing them to fall in toward the Sun. About ten comets per year find their way into the solar system in this way, and some get trapped by the gravity of the planets. Studying these comets may provide information about conditions in our solar system long before the Sun was born.

Below:
An ultraviolet image of the comet Kohoutek, taken from the Skylab *space station in 1973.*

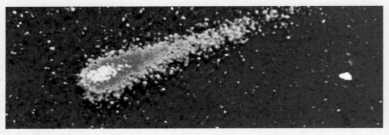

Right:
A comet's illuminated cloud, or coma, becoming brighter as it nears the Sun.

Right:
The probe Giotto *visited Halley's Comet in March 1986. In January 2003* Rosetta *will be launched by an Ariane-5 rocket on an eight-year mission to reach the comet Wirtanen. Once there, the probe will begin a two-year study of the nucleus of the comet.*

CHRISTENING COMETS

A comet is usually named after its discoverer or discoverers. In 1995, for example, U.S. astronomers Alan Hale and Thomas Bopp spotted a small comet independently, and it became known as Hale-Bopp. But perhaps the best-known story behind a comet being named for its discoverer concerns Halley's Comet. For the 18th-century British astronomer Edmond Halley never actually saw it. In 1705, he came across three previous historical appearances of the comet—including that recorded on the Bayeux Tapestry (1066)—and from this calculated that it appeared once every 76 years. Its next showing, he predicted, would be in 1758. He was proved correct, and the comet was named in his honor.

Above:
Comet Hale-Bopp seen from the shuttle orbiting Earth.

Right:
Hale-Bopp's bluish gas tail observed from Earth.

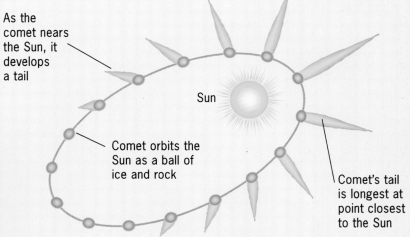

As the comet nears the Sun, it develops a tail

Sun

Comet orbits the Sun as a ball of ice and rock

Comet's tail is longest at point closest to the Sun

Above:
This diagram shows the typical orbital pattern of a comet. Its farthest point from the Sun may be well beyond the planet Pluto.

Below:
A space probe of the future, Stardust will collect dust from the comet Wild 2 in 2004 and return to Earth in 2006.

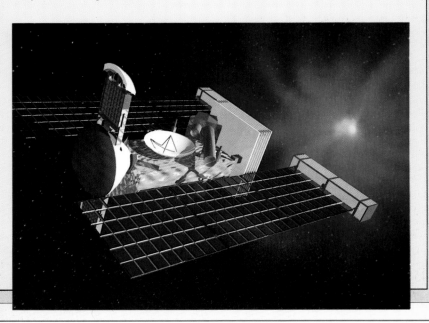

Deep space exploration

See also:
- **The stars and constellations** *p. 78*
- **The Milky Way** *p. 80*

PIONEER PLAQUE

The spacecraft *Pioneer 10* carries a special message to any intelligent life-form in the universe that may come across it. The 6 inch by 9 inch (15 cm by 23 cm) plaque (below) is etched with a diagram of a male and female, and shows their height relative to the size of the spacecraft. The male figure has his hand raised in greeting, while the path of the spacecraft and the solar system is shown in relation to the position of 14 pulsars (neutron stars).

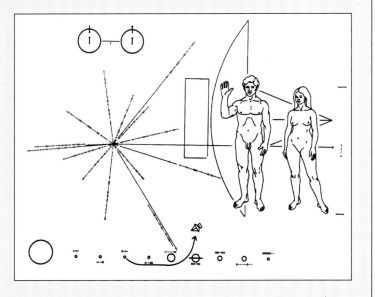

WHAT TRULY LIES beyond our solar system? There are five craft now in space that are destined to voyage indefinitely across the vast void of the universe. Unfortunately, they will be unable to communicate with Earth, and compared to the science-fiction vision of space travel, the progress of *Pioneers 10* and *11* and *Voyagers 1* and *2* will be painfully slow. Though *Voyager 2* now travels at 6 miles per second (10 kps), for example, it will take it over 350,000 years to reach one of our nearest stars, Sirius. It seems high-speed interstellar travel must remain a dream for now.

Above:
Pioneer 11 is launched aboard an Atlas Centaur *rocket on April 6, 1973.*

Right:
Outer space, where the craft will drift.

Above:
One of the Pioneer *spacecraft. Both weighed 569 pounds (258 kg) and were equipped to transmit signals.*

SPACECRAFT FACTS

Pioneer 10
Launched: March 3, 1972
Heading: the star Aldebaran
Arrival time: two million years
Pioneer 11
Launched: April 6, 1973
Heading: the constellation Aquila
Arrival time: four million years
Voyager 1
Launched: September 5, 1977
Heading: the constellation Camelopardus
Arrival time: 400,000 years
Voyager 2
Launched: August 20, 1977
Heading: Sirius
Arrival time: 358,000 years
Microwave Anistropy Probe (MAP)
Launched: June 30, 2001
Heading: the oldest light in the universe

EARTH SAMPLE

Each *Voyager* carries examples of our planet's life and culture in case it is intercepted one day. A 12 inch (30 cm) gold-plated phonograph disc (right), together with a needle and playing instructions, has sounds of thunder, birds, whales, volcanoes erupting, and even human laughter recorded on it. The disc also contains 90 minutes of music, 115 analog pictures, and greetings in 60 languages.

Right:
A Voyager. Each Voyager weighed 1,819 pounds (825 kg) and was launched aboard a Titan IIIE-Centaur booster. Voyager I (far right) was launched on September 5, 1977.

Below:
The flight paths of the deep space missions. Voyager 2 is set to pass within 0.8 light-years of Sirius (below inset), the brightest star in the night sky.

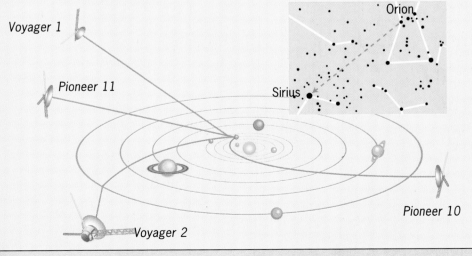

Voyager 1

Pioneer 11

Orion

Sirius

Pioneer 10

Voyager 2

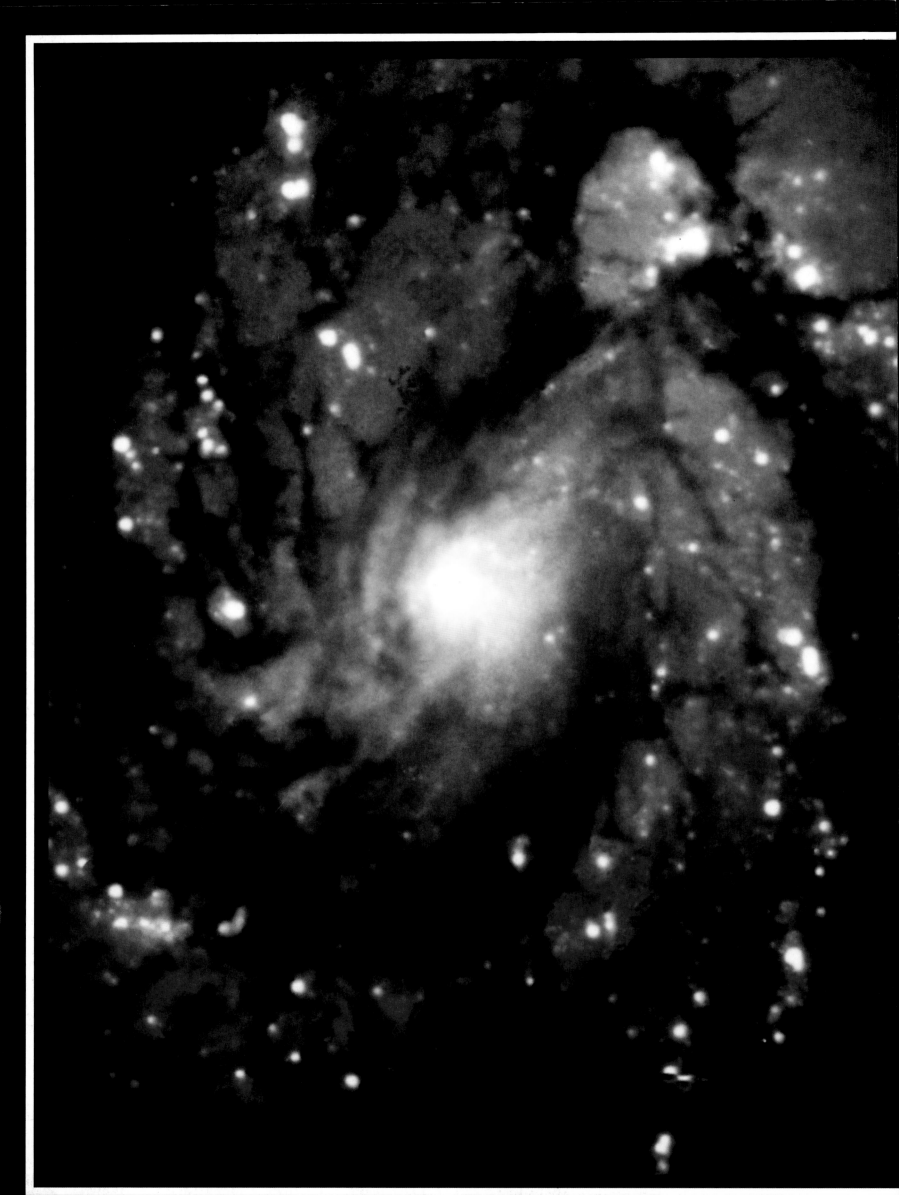

To the stars and beyond

The universe is so vast that it almost defies comprehension. The distance traveled by light in a year is immense, yet deep-voyaging spacecraft have observed parts of the universe as much as 18 billion light-years away. Since travel to the stars is unlikely in the foreseeable future, telescopes must remain our principal source of knowledge. Powerful modern telescopes based on Earth have been joined by "eyes in the sky"—by the *Hubble Space Telescope* and other space observatories. Such instruments have enabled astronomers to look into the very depths of space and pinpoint distant galaxies, supernovae, red giants, black holes, and neutron stars—discoveries that have led to credible theories about the overall structure, origins, and possible fate of the universe. The many mysteries that remain must be puzzled over by future generations of observers.

Above:
A companion galaxy to our own galaxy, the Milky Way.

Above:
The International Space Station—*fully operational in 2004.*

The Hubble Space Telescope

See also:
- **Types of satellites** *p. 20*
- **The space shuttle** *p. 24*

MIRROR ERROR

The *Hubble Space Telescope*'s 8.2-foot (2.5 m) diameter "primary reflecting" mirror needs to be amazingly smooth to register images from deep space. Nonetheless, a slight error was made in the manufacture (left) and the telescope's first images were blurred. A preplanned servicing mission to the telescope by the space shuttle in 1993 thus became a repair mission. A new instrument resembling a pair of glasses was fitted to the *Hubble* to correct its "vision."

Below:
The Hubble Space Telescope *before launch—42.6 feet (13 m) long and powered with two huge solar arrays.*

EARTH-BASED TELESCOPES must cope with a certain amount of distortion or blurring of images caused by our planet's atmosphere. For a telescope to peer into the very depths of space, it needed to be sited beyond our atmosphere in Earth's orbit. The answer was the *Hubble Space Telescope*, deployed by the space shuttle in 1990. Named after a famous 20th-century astronomer, the *Hubble* can "look" 14 billion light-years into the universe—farther than any instrument except an X-ray satellite. But for sheer clarity, the telescope's images are unequaled. They have given us a new sense of the vastness of the universe.

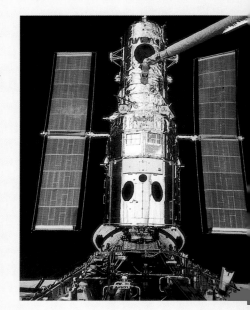

Above and below:
Views of the Hubble *in orbit during the space shuttle's first servicing and repair mission to the telescope in 1993.*

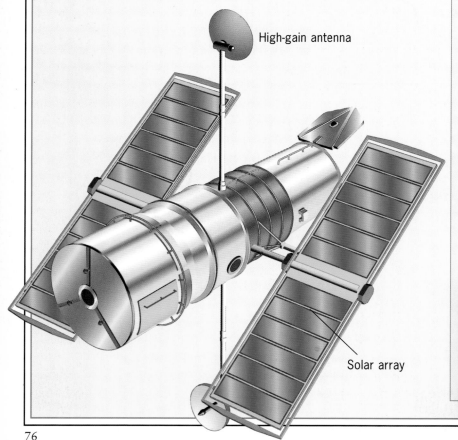

High-gain antenna

Solar array

STRONG VIEWING

With its ability to see 10 times farther than Earth-based telescopes, the *Hubble* (right) is a great advance in astronomy. The six gyroscopes on the instrument can point it with such accuracy that it can "see" the equivalent of a small coin 398 miles (640 km) away.

Below:
In the course of its low orbit, the Hubble *operates best in the Earth's shadow.*

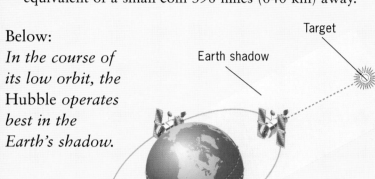

Target

Earth shadow

The *Hubble*

Hubble orbit path

The Sun

Above and right:
Astronauts at work on the Hubble *during the shuttle's second servicing mission to the telescope in 1997.*

Stars and constellations

See also:
- **The Milky Way** p. 80
- **Galaxies** p. 82

BIRTH OF A SUPERNOVA

A supernova is an enormous explosion that occurs when a star burns up all its fuel and its core collapses. The star Eta Carinae (below), believed

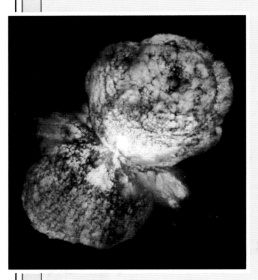

to be 100 times bigger than the Sun, is a supernova in the making. Hidden by dust and gas clouds, it is expanding at 311 miles (500 km) per second. When it eventually explodes, it will shine more brightly than Venus in the night sky.

WHEN THE MATTER in a cloud of gas and dust, or nebula, combines through gravity, great temperatures are created—and a star is born. In a star's early life it is large and bright. Then, as it reaches maturity, it shrinks to some extent and becomes hotter. Finally, when it begins to run out of fuel, it again expands until it may become a red giant 100 times its original size. Some of the largest red giants eventually explode, creating supernovae, which in turn leave behind either black holes or tiny neutron stars.

Below:
The Eagle Nebula, where stars are born with light-year-long "columns."

Above:
The "hourglass" shape of a planetary nebula 8,000 light-years away, created by a dying star's gradual ejection of matter.

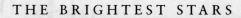

THE BRIGHTEST STARS

Name	Constellation	Magnitude	Distance
1. Sirius	alpha Canis Major	-1.46	8.6 light-years
2. Canopus	alpha Carinae	-0.72	110
3. (Unnamed)	alpha Centauri	-0.01	4.37
4. Arcturus	alpha Bootis	-0.04	36
5. Vega	alpha Lyrae	0.03	26
6. Capella	alpha Aurigae	0.08	45
7. Rigel	beta Orionis	0.12	850
8. Procyon	alpha Canis Minoris	0.8	11.4
9. Achemar	alpha Erindani	0.46	127
10. Hadar	beta Centauri	0.66	520
11. Betelgeuse	alpha Orionis	0.70	650
12. Altair	alpha Aquilae	0.77	16

ORION AND BETELGEUSE

The constellation Orion dominates the winter sky in the northern hemisphere. At its center is the Orion Nebula, where some 700 recently formed stars have been detected. At the top left of the constellation is the aging star Betelgeuse (right), photographed here by the *Hubble Space Telescope*. About 650 light-years away and with a diameter of about 311 million miles (500 million km), the death throes of Betelgeuse could swallow up the solar system as far as Jupiter.

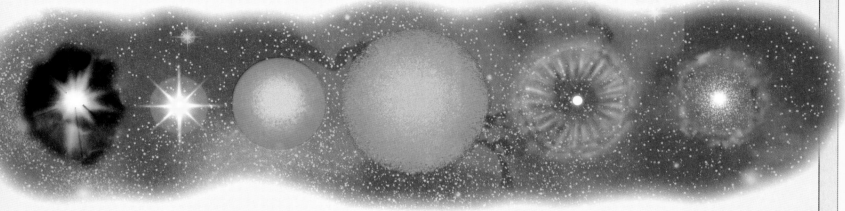

Above:
A nebula collapses under gravity. High temperatures and clusters of young stars are created.

Above:
As they run out of fuel, stars cool and expand to become red giants 10 to 100 times their original size. Larger stars become red supergiants over 500 times the size of the Sun.

Above:
Red supergiants eventually give way to their enormous gravity, collapse at the core, and explode, creating an extremely bright, though temporary, supernova.

Above:
A neutron star left by a supernova. A few miles (km) wide, it has the mass of 100 of our Suns.

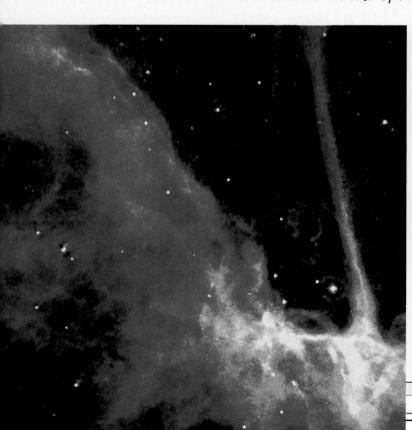

Left:
The blast wave from a supernova that occurred 15,000 years ago. The stars in the constellation Leo, the Lion (above).

The Milky Way

See also:
- **Stars and constellations** p. 78
- **Galaxies** p. 82

THE MILKY WAY is the name commonly given to a great band of remote stars sometimes seen in the night sky. In fact this "band" is toward the core of our spiral-shaped galaxy, and we are looking from a position in one of the spiral's four outer arms. Though by no means the biggest galaxy in the known universe, the Milky Way is vast. Measuring 150,000 light-years across (compared to our solar system's mere 7,452 million miles (12,000 million km), it contains some 200 billion stars, and light from the Sun takes 28,000 light-years to reach its center. The 5,000 stars that can be seen relatively clearly from Earth are those situated very close to our solar system in the same or adjacent spiral arm.

Below:
The Milky Way in an image derived from radio astronomy. It clearly shows the spiral structure and principal arms.

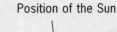

Position of the Sun

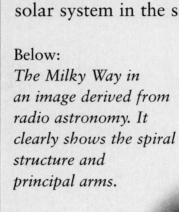

Perseus

Orion

Sagittarius arm

Centaurus arm

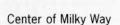

Center of Milky Way

Position of the Sun

Top:
Part of the Milky Way in the region of the Scorpio and Sagittarius constellations. Viewed from the side (above), the Milky Way would resemble a disk with a bulging center.

MILKY WAY ARMS

Our local star, the Sun, is situated in the Orion arm of the Milky Way and contains stars that form constellations such as Orion and Cygnus. The spiral arm adjacent to the Milky Way, the Perseus outer arm, contains stars we can see in the northern hemisphere constellations of Perseus, Cassiopeia, and Gemini. The stars in the Sagittarius arm are mostly seen in the southern hemisphere, as are those in the Centaurus arm.

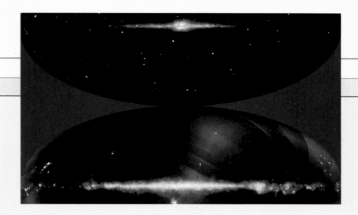

Right:
The view from the spacecraft COBE, reconstructing what the Milky Way looks like from the outside.

Left:
A giant cloud in the Eagle Nebula —a nursery where stars are born.

HUBBLE VISION

Astronomers have been given close views of many parts of the Milky Way by space observatories. The most famous of these is the *Hubble Space Telescope*, placed in Earth's orbit in 1990. Spectacular images it has returned include those of a region of hot young stars, the Orion Nebula, and the Large Magellanic Cloud

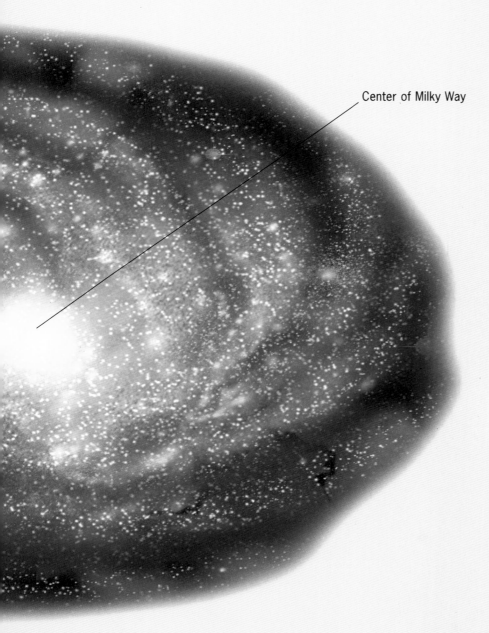

Center of Milky Way

The nebula NGC 2440, a region of cold gas and dust. Here stars are born when gravity pulls this material together into a "ball" that grows hotter and hotter until nuclear fusion occurs. The single star in the center is thought to have a temperature of 360,000 °F (200,000 °C), making it one of the hottest stars known.

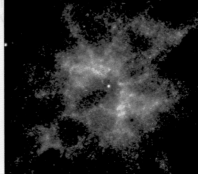

HOTTEST STARS

The Orion Nebula is the largest known stellar nursery in our galaxy. Some 1,500 light-years away and 100 light-years across, the nebula can be seen with the naked eye as the middle of the three "stars" in the "sword" of the constellation Orion. The "sword" itself hangs from a "belt" of three other stars.

ORION NEBULA

The Milky Way has two very small companion galaxies, the Large and Small Magellanic Clouds. These are often clearly visible in the skies of the southern hemisphere. This *Hubble Telescope* image shows 10,000 stars in part of the Large Magellanic Cloud, which covers an area 130 light-years across.

MAGELLANIC CLOUDS

Galaxies

See also:
- **The Milky Way** *p. 80*
- **The universe** *p. 84*

THE CARTWHEEL

Inside the Cartwheel Galaxy 500 million light-years away (below), there is a spiral-like feature in the faint arms or spokes between the outer ring and the bull's-eye-shaped nucleus. That spiral alone contains several billions of new stars and is much bigger than our galaxy. Supernova explosions seem to have broken the spiral's structure in one area and formed a giant bubble of hot gas.

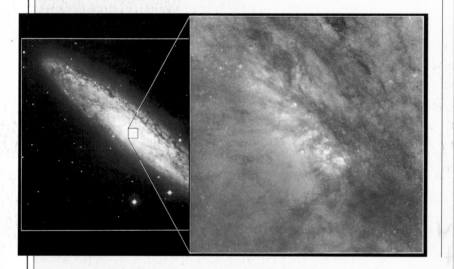

GALAXIES ARE HUGE collections of stars, gas, and dust held together by gravitational force. In the known universe there are four main types of galaxies—spiral, barred spiral, elliptical, and irregular. Our Milky Way is a classic example of a spiral galaxy, having "arms" that lead off a central disc. Elliptical galaxies, by contrast, resemble the core of a spiral galaxy without the arms. Finally, irregular galaxies have little structure or shape. Examples of irregular galaxies include the Milky Way's companions, the Large and Small Magellanic Clouds.

Above:
A "barred" spiral galaxy, NGC 1365, in an image taken by the Anglo-Australian Telescope.

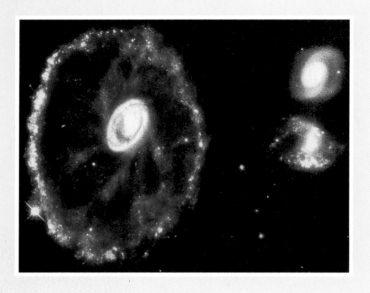

Above:
Focusing on a section within the Andromeda galaxy, a member of our local group of galaxies 2.2 million light-years away, the Hubble Space Telescope *reveals billions of stars.*

GALAXY DATA

The Milky Way is one of a small group of about 30 "local galaxies." The closest that have been named are:
- The Small and Large Magellanic Clouds: small galaxies that orbit the Milky Way about 160,000 light-years distant
- Sculptor: a dwarf galaxy in the constellation of Sculptor seen from the southern hemisphere
- Andromeda: about twice as large as the Milky Way and 2.2 million light-years distant
- Pinwheel Galaxy: the biggest and brightest in the local group and almost face-on to our own galaxy

Right:

The four types of galaxies. The elliptical galaxy (1), which has a dense nucleus like a spiral galaxy. The two types of spiral galaxies, the classic spiral (2), and the barred spiral (3), which has "straightened" arms. Lastly, the irregular galaxy (4).

GALACTIC COLLISION

Galaxies are traveling through space at 19,884,800 miles per second (3.2 million km per second) and it is inevitable that there will be collisions. When two galaxies collide, forces of immense proportions are released. Like a huge boulder thrown into a lake, great ripples of energy flow out at 198,720 miles per hour (320,000 kph), pushing hot glowing gas in front of them. Such collisions also result in the almost instant birth of thousands of stars.

Left:

A Hubble Space Telescope image of a collision between two galaxies, when thousands of star clusters are created in a giant burst of energy.

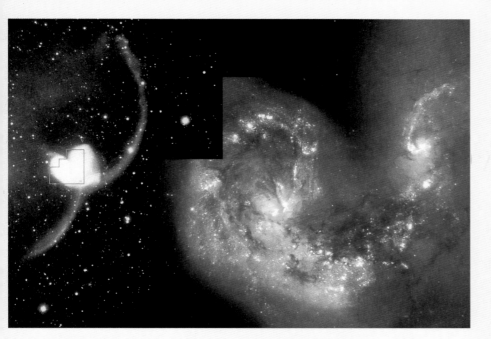

Left:

The Large Magellanic Cloud, an irregular galaxy, makes a vivid contrast with the shapeliness of M100 (right) and M65 (below), which are both spiral galaxies.

The universe

See also:
- **Galaxies** *p. 82*
- **Mysteries of the universe** *p. 86*

ENDLESS CYCLE

With galaxies 13 billion light-years away (below), the sheer immensity of the universe baffles understanding. And it appears to be expanding. Its "end" may then come about through an infinite energy loss in an infinite void. Alternatively, it is possible that expansion will be reversed by the pull of gravity, and that all matter will collapse once again into a single, superdense particle. After that, another universe could be born in another Big Bang—a process that could be repeated forever.

Below:
A COBE satellite image revealing radiation from the Big Bang, the blue and pink ripples representing what we see today as clusters of galaxies.

H OW THE UNIVERSE BEGAN has long been a source of debate among astronomers and physicists. Despite this—and though final proof is perhaps impossible to find—many now accept Edwin Hubble's "Big Bang" theory of its origins. According to this view, all matter and even space itself was once compressed into nothingness. The incredible temperatures and pressures that built up resulted in a huge explosion, which in turn propelled a seething mass of radiation and particles outward. This mass became cooler as it expanded and, with the aid of gravity, galaxies, stars, and planets were formed. But if this theory is true, what existed before the Big Bang?

Above:
The Cosmic Background Explorer, COBE, satellite detected heat radiation left over after the Big Bang.

Above:
An impression of the Big Bang, when matter, time, and space exploded some 18,000 million years ago.

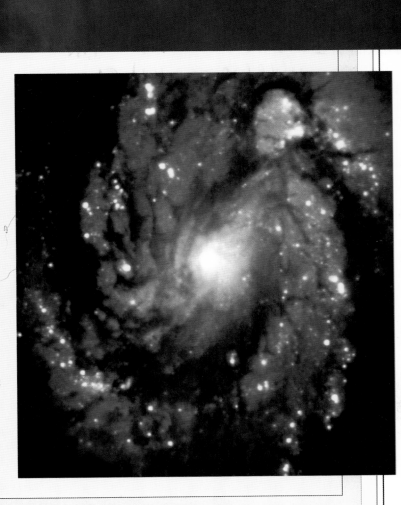

RECEDING GALAXIES

If the expanding universe is thought of as a balloon, and the galaxies as spots on it, then we can see how every galaxy moves away from every other as the balloon inflates. Edwin Hubble discovered that galaxies "recede" from Earth at speeds that increase in proportion to their distance—the farther the galaxy, the faster it recedes. Galaxies recede at a rate of 10 miles per second (16 kps) for every light-year of distance. Thus the Andromeda galaxy is moving toward us at 68 miles per second (110 kps)!

Above:

A group of galaxies flying apart at amazing speeds—evidence that the universe is expanding.

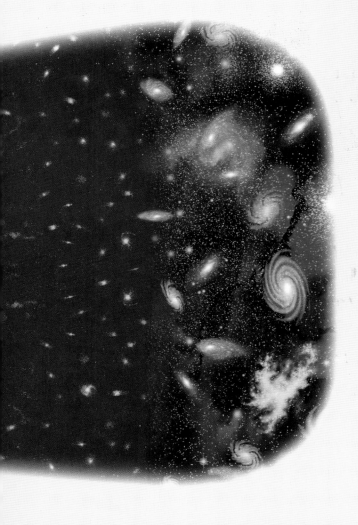

Above:
A tiny region of the universe where stars are being born from a swirling mass of gas and dust.

Right:
A Hubble Space Telescope image of the core of M100, a spiral galaxy lying in the constellation of Virgo.

Mysteries of the universe

See also:
- **Stars and constellations** *p. 78*
- **The universe** *p. 84*

GRAVITATIONAL LENS

The great physicist Albert Einstein saw the universe as a space-time "continuum" in which mass curves space and so creates the gravitational field. Proof of his theory lies in a phenomenon known as gravitational lens. When a galaxy or other massive object in space is observed together with a more distant object, the gravitational field of the large, near object bends the distant object's light. In the telescope the latter appears as a double or even a multiple image (below).

HOW OLD IS THE universe? Is it slowly dying as its energy is used up, or is it continuously expanding? Such great questions remain largely unanswered, though other secrets, such as the true nature of pulsars, quasars, and black holes, have been partially revealed. Pulsars are neutron stars spinning so fast that they send out radio waves, rather as a lighthouse flashes its beacon. Quasars, though starlike in some ways, are thought to be the cores of large galaxies in which there is violent activity. And black holes are formed by the gravitational collapse of stars at the end of their lives.

Above:
An impression of a quasar—a "rod-shaped" emission of energy at the heart of a galaxy.

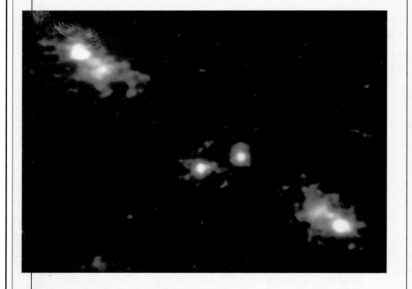

Above and right:
Hubble Space Telescope *images of possible quasars, emitting a huge amount of light and energy.*

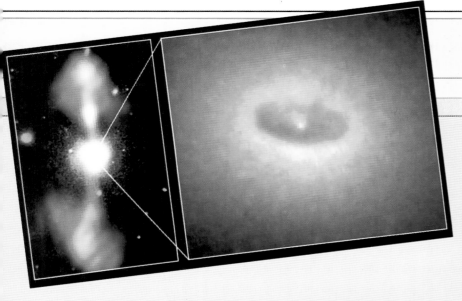

BLACK HOLES

After a supernova, matter may be compressed so much that space and time become distorted. The result is a black hole, whose gravitational pull is so great it sucks in all available material and light. While the whole event may take an instant by "star time," by Earth time it may have occurred over millions of years. Astonishingly, if a star 10 times the size of the Sun were sucked into a black hole, the opening would only be 18.6 miles (30 km) wide (below).

Material is sucked toward a black hole by huge gravitational force

Material creates a vortex as it is sucked into the black hole

Above:
These cometlike objects in a nebula remain a mystery. However, we now have an idea of what a black hole seen by an observer on a nearby planet would look like (left).

Right:
The Space Infra-Red Telescope Facility, designed to look farther into space with infrared than ever before, was launched by NASA in 2000.

The future in space

See also:
- **Space stations** *p. 26*
- **Living and working in space** *p. 28*

POWER STATIONS IN ORBIT

Scientists have designed a huge solar power station (below) to be assembled in Earth's orbit. It would function as follows. Solar arrays would convert the Sun's energy into electricity, which would then be transmitted in microwaves down to Earth. Antennae up to 5 miles (8 km) wide would receive the microwaves, which in turn would be transformed into direct current. Such a space power station could provide up to 10,000 megawatts—enough to supply one million homes. It remains to be seen whether such a project would be cost effective.

MEN HAVE LANDED on the Moon, hundreds of satellites orbit the Earth, and spacecraft have explored distant planets. Yet many of the greater ambitions of space scientists, such as manned space bases on the Moon or on Mars, have yet to be realized. For not only is space exploration extremely expensive, new technology has been more difficult to develop than was at first anticipated. Still, there exists a blueprint for a modest manned mission to Mars, while a manned space plane, the *VentureStar*, could be flying by 2005 and making trips to the *International Space Station.*

Above:
The VentureStar *as it will look in orbit and docked to the* International Space Station *(below).*

LUNAR TELESCOPES

Another future space project now being put forward is an
astronomical observatory on the Moon (right). Unlike Earth-
based telescopes, telescopes on the Moon would not have to
cope with atmospheric blurring. They could also be much
larger than orbital instruments like the *Hubble* and have a
stable platform on which to operate. But assembling such an
observatory would not be an easy task. Astronauts would
have to undertake numerous, highly expensive
missions to the lunar site.

Right:
*Impressions of two
possible future
missions—manned
exploration of one of
the Martian moons
and the planet's Valles
Marineris canyon.*

Below:
*An impression of a
NASA plan to establish
an initial Martian base
within the next 10 years
using* International
Space Station-*type
pressurized modules.*

Tables

MILESTONES IN ASTRONOMY

3000 B.C.
The Egyptians develop the science of astronomy with a basic understanding of the movements and positions of the stars.

2000 B.C.
Passages in the Bible indicate that the Earth is round and suspended in space.

600 B.C.
Greek philosopher Thales of Miletus states his opinion that the Earth is round.

300 B.C.
Greek astronomer Aristarchus suggests that the Earth orbits the Sun.

130 B.C.
Greek astronomer Hipparcos compiles a catalog of 1,000 stars in the night sky.

140 A.D.
Ptolemy of Alexandria produces an encyclopedia of astronomy.

1054 A.D.
Chinese astronomers record the supernova that created the Crab Nebula, visible today in the constellation of Taurus.

1543 A.D.
Copernicus states that the Earth is a planet orbiting the Sun and attempts to discredit the age-old idea that the Earth is at the center of the universe.

1609 A.D.
Galileo makes the first-known astronomical observations using a telescope. He studies craters on the Moon, the moons of Saturn, and the phases of Venus.

1655 A.D.
Huygens discovers the moon Titan orbiting Saturn and describes the true nature of the planet's ring system.

1666 A.D.
Isaac Newton establishes the basic laws of gravity and two years later builds the first reflecting telescope.

1705 A.D.
Edmund Halley predicts that a comet will appear in the skies in 1758. It did, and the comet was named after him.

1781 A.D.
British astronomer William Herschel discovers the first planet using a telescope. The planet is called Uranus.

1801 A.D.
Giuseppe Piazzi discovers the asteroid Ceres.

1846 A.D.
Johann Galle, following the predictions of earlier astronomers, discovers the planet Neptune.

1925 A.D.
Edwin Hubble demonstrates that the universe is expanding, leading to the Big Bang theory of creation.

1930 A.D.
Pluto is discovered by Clyde Tombaugh.

1931 A.D.
Radio waves are detected coming from outer space. The first radio telescope is built six years later.

1955 A.D.
The UK's Jodrell Bank radio telescope is built.

1990 A.D.
The *Hubble Space Telescope* is launched into orbit. Its spectacular images revolutionize astronomy, enabling mankind to see 14 million light-years into the universe.

2001 A.D.
MAP probe launched to search for the oldest light in the universe and learn about Big Bang.

FIRSTS IN PLANETARY EXPLORATION

September 1959 – first moon impact by Soviet Union's *Luna 2*
October 1959 – first pictures of lunar far side from *Luna 3*
December 1962 – first flyby of Venus by U.S. *Mariner 2*
July 1964 – first close-up images of the Moon from *Ranger 7*
July 1965 – first successful flyby of Mars with first images from *Mariner 4*
January 1966 – first "soft" lunar landing and surface images from *Luna 9*
April 1966 – first lunar orbiter is *Luna 10*
October 1967 – first successful exploration of Venus atmosphere by *Venera 4*
September 1968 – first spacecraft to fly round the Moon and return to Earth, *Zond 5*
September 1970 – first return of sample from Moon by unmanned craft, *Luna 16*
November 1970 – first lunar rover is *Lunakhod 1*
December 1970 – first successful Venus landing by *Venera 7*
November 1971 – first Mars orbiter is *Mariner 9*
December 1973 – first craft to explore Jupiter, *Pioneer 10*
March 1974 – first flyby of Mercury by *Mariner 10*
October 1975 – first craft to orbit Venus and return surface pictures, *Venera 9*
July 1976 – first soft landing on Mars by *Viking 1*
September 1979 – first exploration of Saturn by *Pioneer 11*
October 1983 – first Venus radar-mapper, *Venera 15*
January 1986 – first exploration of Uranus by *Voyager 2*
March 1986 – first encounter with coma of a comet (Halley) by *Giotto*
August 1989 – first exploration of Neptune by *Voyager 2*
October 1991 – first flyby of an asteroid, Gaspra, by *Galileo*
December 1995 – first craft to enter Jupiter's atmosphere, *Galileo* probe
December 1995 – first Jupiter orbiter, *Galileo*

FIRSTS IN SATELLITE TECHNOLOGY

October 1957 – first artificial Earth satellite, Soviet Union's *Sputnik 1*
November 1957 – first living being (the dog Laika) in orbit aboard *Sputnik 2*
January 1958 – first American satellite, first science satellite, *Explorer 1*
December 1958 – first experimental communications satellite, *Score 1*
April 1959 – first military spy satellite, *Discoverer 2*
April 1960 – first weather satellite, *Tiros 1*
April 1960 – first navigation satellite, *Transit 1B*
August 1960 – first recovery of craft from orbit, *Discoverer 13*
August 1960 – first recovery of living creatures (two dogs) from orbit, *Sputnik 5*
July 1962 – first commercial communications satellite, *Telstar 1*
July 1963 – first geostationary orbiting communications satellite, *Syncom 2*
April 1966 – first astronomical satellite, *OAO 1*
December 1966 – first French satellite launch, *A1*
October 1967 – first automatic, unmanned docking in orbit, *Cosmos 186-188*
February 1970 – first Japanese satellite launch, *Ohsumi*
April 1970 – first Chinese satellite launch, *Tungfanghung*
October 1971 – first British satellite launch, *Prospero*
July 1972 – first Earth resources remote sensing satellite, *Landsat 1*
February 1976 – first maritime mobile communications satellite, *Marisat 1*
July 1980 – first Indian satellite launch, *Rohini*
June 1981 – first European operational satellite launch by Ariane, *Meteosat 2*
April 1984 – first satellite to be captured, repaired, and redeployed, *SMM 1*
May 1984 – first fully commercial satellite launch, *Spacenet 1*

November 1984 – first satellite capture and return to Earth, *Palapa* and *Westar*

February 1986 – first privately operated, commercial remote sensing craft, *Spot 1*

September 1988 – first Israeli satellite launch, *Ofeq 1*

November 1988 – first unmanned space shuttle launch and landing by Russia, *Buran*

December 1988 – first privately operated, commercial TV satellite, *Astra 1A*

April 1990 – first optical telescope in orbit, *Hubble Space Telescope*

October 2000 – first successful commercial sea launch of the *Thuraya-1* communication satellite

MANNED SPACE RECORDS
(including Russia and the U.S. with the highest ranking other country)

Most spaceflights
MEN
6: John Young (U.S.)
  Story Musgrave
  Franklin Chang Diaz
5: V. Dzhanibekov (Russia)
  Gennadi Strekalov
  Anatoli Solovyov
3: Ulf Merbold (Germany)
  Claude Nicollier (Switzerland)
WOMEN
5: Shannon Lucid (U.S.)
2: Svetlana Savitskaya (Russia)
  Yelena Kondakova (Russia)

Longest spaceflights
MEN
437 days: Valeri Poliakov (Russia)
188 days: Thomas Reiter (Germany)
144 days: Michael Foale (U.S.)
WOMEN
188 days: Shannon Lucid (U.S.)
169 days: Yelena Kondakova (Russia)
15 days: Claudie Andre-Deshays (France)

Most experienced space travelers
MEN
678 days, 2 flights: Valeri Poliakov (Russia)

208 days, 2 flight: Jean-Pierre Haignere (France)
170 days, 4 flights: Michael Foale (U.S.)
WOMEN
223 days, 5 flights: Shannon Lucid (U.S.)
178 days, 2 flights: Yelena Kondakova (Russia)
15 days, 1 flight: Claudie Andre-Deshays (France)

Oldest space travelers
MEN
77 yrs: John Glenn (U.S.)
54 yrs: Gennadi Strekalov (Russia)
53 yrs: Ulf Merbold (Germany)
WOMEN
53 yrs: Shannon Lucid (U.S.)
46 yrs: Roberta Bondar (Canada)
37 yrs: Yelena Kondakova (Russia)

Youngest space travelers
MEN
25 yrs: Gherman Titov (Russia)
28 yrs: Dumitriu Prunariu (Romania)
32 yrs: Eugene Cernan (U.S.)
WOMEN
26 yrs: Valentina Tereshkova (Russia)
27 yrs: Helen Sharman (UK)
32 yrs: Sally Ride (U.S.)

Longest spacewalks
MEN
8hr 29min: Pierre Thuot, Rick Hieb, Tom Akers (U.S.), Earth orbit
7hr 37min: Eugene Cernan, Jack Schmitt (U.S.), Moon walk
7hr 16min: Anatoli Solovyov, Alexander Budarin (Russia), Earth orbit
5hr 57min: Jean-Loup Chrétien (France), Earth orbit
1hr 24min: Ken Mattingly (U.S.), trans-Earth
1hr 22min: Bruce McCandless (U.S.), untethered
WOMEN
7hr 49min: Kathryn Thornton (U.S.), Earth orbit
3hr 55min: Svetlana Savitskaya (Russia), Earth orbit

Most spacewalks
MEN

14: Anatoli Solovyov (Russia)
7: Jerry Ross (U.S.)
2: Thomas Reiter (Germany)

WOMEN
3: Kathryn Thornton (U.S.)

Most experienced spacewalkers
MEN
71hr, 14 Earth orbits: Anatoli Solovyov (Russia)
44hr, 7 Earth orbits: Jerry Ross (U.S.)
8hr, 2 Earth orbits: Thomas Reiter (Germany)
WOMEN
21hr 15min, 3 Earth orbits: Kathryn Thornton (U.S.)

FAMOUS FIRSTS IN MANNED SPACEFLIGHT

First in space: Yuri Gagarin (USSR) Apr. 12, 1961
First American in space: Alan Shepard May 5, 1961
First American in orbit: John Glenn Feb. 20, 1962
First woman in space: Valentina Tereshkova (USSR) Jun. 16, 1963
First non-US, non-Soviet in space: Vladimir Remek, (Czechoslovakia) Mar. 2, 1978
First to make two flights: Gus Grissom (U.S.) Mar. 25, 1965
First to make three flights: Wally Schirra (U.S.) Oct. 11, 1968
First to make four flights: James Lovell (U.S.) Apr. 11, 1970
First to make five flights: John Young (U.S.) Apr. 12, 1981
First to make six flights: John Young (U.S.) Nov. 28, 1983
First to walk in space: Alexei Leonov (USSR) Mar. 18, 1965
First to walk in space independently using MMU: Bruce McCandless (U.S.) Feb. 3, 1984
First woman spacewalker: Svetlana Savitskaya (USSR) July 17, 1984
First male-female spacewalk: Vladimir Dzanibekov, Svetlana Savitskaya (USSR) July 17, 1984
First spacewalk between Moon and Earth: Alfred Worden (U.S.) July 26, 1971
First docking: *Gemini 8* (U.S.) Mar. 16, 1966

First two-crew flight: *Voskhod 2* (USSR) Mar. 18, 1965
First three-crew flight: *Voskhod 1* (USSR) Oct. 12, 1964
First four-crew flight: *STS 5* (U.S.) Nov. 11, 1982
First five-crew flight: *STS 7* (U.S.) Jun. 18, 1983
First six-crew flight: *STS 9* (U.S.) Nov. 28, 1983
First seven-crew flight: *STS 41G* (U.S.) Oct. 5, 1984
First eight-crew flight: *STS 61A* (U.S.) Oct. 30, 1985
First flight to the Moon: *Apollo 8* (U.S.) Dec. 21, 1968
First flight to land on the Moon: *Apollo 11* (U.S.) July 16, 1969
First men on Moon: Neil Armstrong, Buzz Aldrin (U.S.) July 21, 1969

MOONWALK LOG

Apollo 11: N. Armstrong/E. Aldrin, Jul. 21, 1969, 2hr 21min
Apollo 12: C. Conrad/A. Bean, Nov. 19, 1969, 3hr 56min
Apollo 12: C. Conrad/A. Bean, Nov. 20, 1969, 3hr 49min
Apollo 14: A. Shepard/E. Mitchell, Feb. 5, 1971, 4hr 49min
Apollo 14: A. Shepard/E. Mitchell, Feb. 6, 1971, 4hr 35min
Apollo 15: D. Scott/J. Irwin, Jul. 31, 1971, 6hr 14min
Apollo 15: D. Scott/J. Irwin, Aug. 1, 1971, 6hr 55min
Apollo 15: D. Scott/J. Irwin Aug. 2, 1971, 4 hr 27min
Apollo 16: J. Young/C. Duke, Apr. 20, 1972, 7hr 11min
Apollo 16: J. Young/C. Duke, Apr. 21, 1972, 7hr 23min
Apollo 16: J. Young/C. Duke, Apr. 22, 1972, 5hr 40min
Apollo 17: E. Cernan/H. Schmitt, Dec. 12, 1972, 7hr 12min
Apollo 17: E. Cernan/H. Schmitt, Dec. 13, 1972, 7hr 37min
Apollo 17: E. Cernan/H. Schmitt, Dec. 14, 1972, 7hr 16min

Glossary

A

Asteroid
A large rocky object orbiting the Sun, thought to be material left over from the formation of the *solar system*.

Atmosphere
A layer of gas surrounding a *planet*, *star*, or *moon*.

Aurora
Phenomenon caused by impact of solar particles on Earth's upper *atmosphere*.

Axis
The angle at which a *planet*'s north pole points in relation to its *orbit*. For example, the Earth's axis is tilted 23° relative to its orbit of the Sun.

B

Big Bang
An explosion of space, time, and matter about 18 billion years ago—the most popular current theory for the origins of the universe.

Black hole
A region in space where the concentration of matter is so dense that not even light can escape. Black holes are often formed when a dying *star* collapses and sucks material, including light, into itself.

Booster
A term often used for the rocket that propels a *payload* or *module* into space.

C

Comet
A small icy object surrounded by gas and dust that moves in a highly *elliptical orbit* around the Sun. Comets that near the Sun leave spectacular "tails" of gas and dust.

Constellation
A group of stars, often named after a mythological figure or object.

Corona
The high-temperature outer *atmosphere* of the Sun, most clearly visible during a total solar *eclipse*.

Cosmonaut
The name given to a Russian space traveler.

Cosmos
Alternative term for the *universe* and everything in it.

Crater
A circular hollow or depression in the surface of a *planet* or *moon*, usually the result of *meteor* impact.

D

Docking
The joining process of two spacecraft.

E

Eclipse
The passage of one astronomical body in front of another—such as takes place during a total eclipse of the Sun by the Moon.

Ecliptic
The plane of the Earth's orbit around the Sun.

Electromagnetic radiation
The range of radiation, from gamma rays through the *spectrum* of visible light to radio waves.

Elliptical orbit
The noncircular movement of one body in space around another. The planets and comets orbit the Sun in elliptical orbits.

Equator

The imaginary line around the center of the Earth or other *planet*.

F

False-color image
An image with color added by a computer.

G

Galaxy
A huge group of millions of stars held together by *gravity*.

Geostationary orbit
A circular *orbit* around the Earth's equator at a height at which the speed of a *satellite* is exactly the same as the speed of the Earth's rotation. Thus the satellite appears to be stationary in the sky.

Gibbous
Term describing the phase of the Moon between half and full illumination.

Gravitational lens
The phenomenon that occurs when the gravity of a large astronomical body, such as a *galaxy*, "bends" the light from a more distant object and creates double images.

Gravity
The fundamental property of matter, which produces mutual attraction.

H

Hemisphere
Half of a sphere such as a planet or moon.

I

Inclination
The angle at which a *satellite* crosses the equator in its orbit of a planet or moon.

Infrared radiation
Electromagnetic radiation beyond the red end of the visible spectrum.

L

Latitude
A coordinate for determining positions on Earth or another planet north or south of the *equator*.

Light-year
The distance light, traveling at 186,282 miles per hour (299,792 kps), travels in one year. Light from Sirius, the brightest *star* in the night sky, takes eight years to reach Earth.

Local Group
A cluster of about 20 known galaxies including our own, the Milky Way.

M

Magellanic clouds
Two relatively small, irregular galaxies located close to the Milky Way, and visible in the night sky of the Earth's southern hemisphere.

Magnitude
The measure of the relative brightness of a *star* or other object in space.

Mantle
The matter surrounding the core of a *planet*.

Mass
The measure of a body's resistance to acceleration.

Meteor
A relatively small, rocky object that was left over from the formation of the *solar system*. Thousands of meteors orbit the Sun, and many enter and burn up in the Earth's *atmosphere*.

Milky Way
Our *galaxy*. Also—confusingly—the common name for the "cloud band" of stars that can sometimes be seen in the night sky.

Module
A self-contained unit forming part of a spacecraft.

Moon
A small natural object orbiting a *planet*.

N

Nebula
A gaseous and dusty region of space in which stars are formed or which remains when a *star* dies.

Neutron star
The tiny *star* left over when a star runs out of fuel and dies.

O

Orbit
The path followed by an astronomical or natural object under the gravitational sway of another body in space.

P

Payload
The cargo carried into space by a spacecraft or "boosted" into it by a *rocket*.

Phase
The proportion of an illuminated body in space that is visible to an observer.

Photosphere
The visible surface of the Sun—a gaseous layer several hundred miles (km) thick.

Planet
A large body that orbits a *star*.

Pulsar
A rotating *neutron star* that sends out regular pulses of radio signals.

Q

Quasar
The most powerful and energetic object in the universe, thought to be the energetic core of an active *galaxy*.

R

Radiation
The wavelengths in the *electromagnetic spectrum*: infrared, X rays, ultraviolet light, visible light, gamma rays, and radio waves.

Radio telescope
An instrument for detecting and measuring *electromagnetic radiation* of radio frequencies from outer space.

Red giant
A *star* in a late stage of life that has cooled and expanded to perhaps 100 times its original size.

Reentry
The moment when a *satellite* is captured by the Earth's gravity and begins to be pulled back into the *atmosphere* at high speed.

Remote-sensing satellite
A *satellite* equipped with cameras to survey the Earth's resources and environment.

Resolution
The degree of definition and clarity in a photographic image.

Rocket
A space vehicle that is forced through space by jet propulsion and carries its own fuel and oxidizers.

S

Satellites
Objects that *orbit* other objects in space—normally applied to spacecraft orbiting the Earth but also used by astronomers to designate the *moons* of a *planet*.

Solar cells
Small mirrorlike panes of silicon that concentrate the energy of the Sun. Thousands of solar cells are usually carried on the winglike solar arrays of *satellites*.

Solar flare
A brilliant eruption of light from the surface of the Sun.

Solar system
A system of planets and other objects orbiting a *star* such as the Sun.

Solar wind
A continuous outward flow of particles from the Sun's *corona* into space.

Solstice
The moment when the Earth's *axis* is inclined at its maximum toward the Sun.

Spectrum
The range of *electromagnetic radiation* of different *wavelengths*. Also light divided into its wavelengths of different colors.

Spiral galaxy
A type of *galaxy* in which many of the stars and nebulae lie along spiral arms.

Star
A self-luminous body in space that generates nuclear energy within its core.

Suborbital
Term to describe a straight-forward "up and down" spaceflight that does not reach Earth's *orbit*.

Sunspot
A relatively cool dark area on the Sun.

Supernova
The explosion of a massive *star* that blows off its outer atmosphere and temporarily becomes extremely bright.

T

Thermostat
An automatic device for controlling temperature.

Thrust
The force of burnt fuel or exhaust that propels a *rocket* through space.

U

Ultraviolet radiation
Electromagnetic radiation of a shorter wavelength than the blue end of the visible spectrum.

Universe
All the matter, space, and time that exists.

W

Wavelength
The distance between the peaks or troughs of successive radiation waves.

White dwarf
A small, hot, dense star—a faint remnant of a *red giant* that has lost its outer layers of gas as a planetary *nebula*.

X

X rays
Electromagnetic radiation of a shorter wavelength than *ultraviolet radiation*.

Index

Acknowledgments

The publishers wish to thank the following organizations and individuals for providing photographs for use in this publication

l=left, *r*=right, *t*=top, *c*=center, *b*=bottom

3-8 *all* NASA/Genesis Space Photo Library; 8-9 all Heritage Photo Collection (HPC); 10-11 *all* Anglo Australian Observatory/David Malin, except *tr* Arecibo Observatory; 12-13 *all* NASA/Genesis Space Photo Library, except 12, *l* Novosti Photo Library; 14 *t* China Great Wall Industry/Genesis Space Photo Library, *b* Arianespace/Genesis Space Photo Library; 11 *l* Arianespace/Genesis Space Photo Library, tr, ml Lockheed Martin/Genesis Space Photo Library, *b* Boeing, *br* ILS International Launch Services/Genesis Space Photo Library; 16 *tl* Novosti Photo Library, *others* NASA/Genesis Space Photo Library, 17 *cr* Novosti Photo Library, *others* NASA/Genesis Space Photo Library; 18 *l* Aerospatiale/Genesis Space Photo Library, *r* Hughes /Genesis Space Photo Library; 19 *tl, c* NASA Genesis Space Photo Library, *r* Matra Marconi Space/Genesis Space Photo Library, *r* Hughes/Genesis Space Photo Library; 20 *l* NASA/Genesis Space Photo Library, *b* TRW/Genesis Space Photo Library, *r* European Space Agency/Genesis Space Photo Library; 21 *t* NASA/Genesis Space Photo Library; *c* Lockheed Martin/Genesis Space Photo Library, *b* Sea Lauch; 22 *r* Novositi Space Photo Library, *others* NASA/Genesis Space Photo Library; 23 *all* NASA/Genesis Space Photo Library; 24-25 *all* NASA/Genesis Space Photo Library, except 25 *r* NASA/HPL; 26 *tl* Boeing/Genesis Space Photo Library, *tr* Novosti Photo Library; 27 *all* NASA/Genesis Space Photo Library; 28-29 *all* NASA/ Genesis Space Photo Library; 30-31 *all* NASA/Genesis Space Photo Library; 32-33 *all* NASA/Genesis Space Photo Library; 34-35 *bl* ESA/Genesis Space Photo Library, *all others* NASA/Genesis Space Photo Library; 36-37 *all* NASA/ Genesis Space Photo Library; 38 *r* Novosti Photo Library, *l* NASA/Genesis Space Photo Library; 39 *all* NASA/Genesis Space Photo Library; 40-41 *all* NASA/ Genesis Space Photo Library; 40-41 *all* NASA/Genesis Space Photo Library; 42 NASA/Genesis Space Photo Library; 43 *t* Spot Image, *tr, br* National Remote Sensing Centre; *others* NASA/Genesis Space Photo Library; 44-45 *tl* Boeing/Genesis Space Photo Library, *all others* NASA/Genesis Space Photo Library; 46 *l, c, b* Novosti Photo Library, *tr* NASA/Genesis Space Photo Library; 47 *tr* Hughes/Genesis Space Photo Library, *all others* NASA/Genesis Space Photo Library; 48-49 *all* NASA Genesis Space Photo Library; 50 *cl* CNES, *others* NASA/Genesis Space Photo Library, 51 NASA/Genesis Space Photo Library; 52-53 *all* NASA/Genesis Space Photo Library; 54 *c* David Hardy/Astro Art; 55 *tl* NASA/Genesis Space Photo Library; *r* Kitt Peak National Observatory/Galaxy Picture Library, *c* Tswaing Soutpan Meteor Crater company, *b* SpaceDev/Genesis Space Photo Library; 56-57 *all* NASA/Genesis Space Photo Library; 58-59 *all* Genesis Space Photo Library; 60-61 *l* European Space Agency/Genesis Space Photo Library, *all others* NASA/Genesis Space Photo Library; 62-63 *tl, tr* Aerospatiale/Genesis Space Photo Library, *others* NASA/Genesis Space Photo Library; 64-65 *tl* HPC, *others* NASA/Genesis Space Photo Library; 66-67 *all* NASA/Genesis Space Photo Library; 68-69 *tr* David Hardy/Astro Art, *others* NASA/Genesis Space Photo Library; 70-71 *c* European Space Agency/Genesis Space Photo Library, *tr* Robin Scagell/Galaxy Picture Library, *others* NASA/ Genesis Space Photo Library; 72-73 *all* NASA/Genesis Space Photo Library; 74-75 *all* NASA/Genesis Space Photo Library; 76-77 *tl* Lockheed Martin/Genesis Space Photo Library, *others* NASA/Genesis Space Photo Library; 78-79 *all* NASA/Genesis Space Photo Library; 80-81 *tl* Robin Scagell/Galaxy Picture Library, *others* NASA/Genesis Space Photo Library; 82 *l* NASA/Genesis Space photo Library, *r* Anglo Australian Observatory/David Mailin, *c* Rutherford Appleton Laboratory; 83 *all* NASA/Genesis Space Photo Library; 84-85 NASA/Genesis Space Photo Library; Pages 86-87 *c* European Space Agency/Genesis Space Photo Library, *others* NASA/Genesis Space Photo Library; 88 *l* Boeing/Genesis Space Photo Library, *r, b* Lockheed Martin/Genesis Space Photo Library; 89 *t* European Space Agency/Genesis Space Photo Library, *others* NASA/ Genesis Space Photo Library

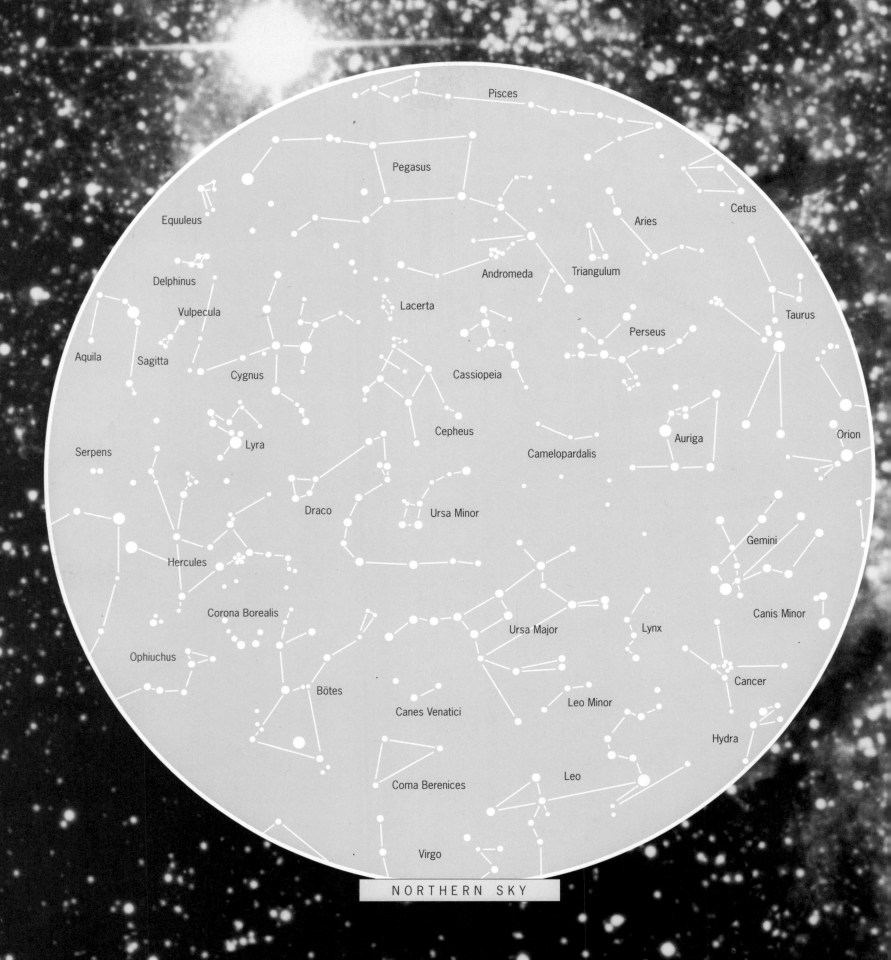

NORTHERN SKY